Wave Motion as Inquiry

Fernando Espinoza

Wave Motion as Inquiry

The Physics and Applications of Light and Sound

 Springer

Fernando Espinoza
Hofstra University and SUNY College at Old Westbury
Long Island, NY, USA

ISBN 978-3-319-45756-7   ISBN 978-3-319-45758-1   (eBook)
DOI 10.1007/978-3-319-45758-1

Library of Congress Control Number: 2016953733

Printed on acid-free paper

This Springer imprint is published by Springer Nature
The registered company is Springer International Publishing AG
The registered company address is: Gewerbestrasse 11, 6330 Cham, Switzerland

Preface

This book is written for an audience that is very diverse in its learning styles, and my objective is to expose the readers to different and unique ways the basics of wave motion can be studied in an academic setting. Why such a title? There have been many books written about waves, and quite a few are sufficiently successful in covering and teaching a wide range of topics to provide exposure to the physics of wave motion. There are books about waves in general and specifically about light and about sound that convey the basic ideas that all waves follow.

Because most of these books emphasize the coverage of topics from an authoritative perspective, they have neglected the student perspective.

By student perspective I mean one where the ability to apply the ideas is inseparable from being exposed to their basic definitions. From my experiences as a physics student and instructor, the traditional presentation has been that the didactic approach often neglects the context.

In the science education literature researchers have known for some time that from the earliest experiences, humans tend to be better observers when they are interested. Correspondingly, it seems logical to suppose that a learner being provided with a context stands a better chance of understanding the material, rather than being introduced to it without one.

Understanding a concept involves much more than simply memorizing and regurgitating the information. The application of concepts places a learner in a situation where more is required than a simple recollection of information. To this extent, the role of inquiry must be actively incorporated into instructional materials, if one hopes to appeal to that natural need of a context.

Inquiry-based instruction needs to place the learner in a position of relative ignorance, although some guidance needs to be provided; the reason for this guidance is to allow the learner to utilize whatever background information he or she possesses. There are many views on the importance of inquiry, some claiming that it doesn't show substantial gains in the learning of difficult concepts. Many studies, however, have shown a clear advantage for nontraditional students to be engaged in inquiry-based instruction.

Consider an example using light and sound as sources of information to explore and understand the world. Generally speaking, events do not become experiences until there is awareness, and the experiences become more meaningful when there is reflection upon them. Suppose you are seated outdoors and hear a bird singing by emitting short bursts of a high-pitched sound and you want to locate the bird; there will be an interaction between the waves that make up what you see and what you hear. Both types of waves undergo similar processes as they get to you, for the most part; however, you will not experience them all due to various properties, such as the length of the waves, how they propagate, what other objects do to these waves when struck by them, and so on.

If you were an ornithologist, you most likely would know what type of bird it is, since the example is about hearing it but not seeing it. Additionally, you could probably tell whether it is a female or male, how old it is, its size, etc. However, if you were not an expert on bird watching, you would need to rely on the sound to guide your vision if you hope to see it. There are some things you could ascertain about the bird from prior experience, such as its size and even its type. For one thing you can deduce that if the sound is high pitched, the bird is not a crow, which would also make it easier to spot. Based on this, you conclude that the bird is small. If there are several trees having lots of branches and leaves, this makes the task more difficult since you seem to hear the sound coming from various directions. As you continue to listen, you begin to concentrate on an area where you think the sound is coming from based on what you hear; however, if the bird were to stop singing, you would be extremely hard pressed to find it.

You could probably come up with a better example than this to provide a setting; what I have done is to create a scenario and provide an experience. I believe that is the essence of the term inquiry, the provision of opportunities to experience phenomena and to explore them based on what one knows beforehand.

The way in which this book deals with the incorporation of inquiry is by its lack of distinction between theory and practice. As soon as a concept or an idea is introduced, an attempt is made to provide opportunities for exploration. Whether the task is one of concept development or quantitative determinations, to provide opportunities to explore lies at the heart of inquiry-based instruction.

Additionally, the text incorporates laboratory experiences into the introduction of the content; while this has been done before, the types of experiences are both physical and virtual. There is an undeniable benefit to being able to manipulate variables in a setting that does not require extensive preparation and where the data obtained can be processed in real time.

There are many tasks involving simulations, predominantly using one of the very best sources publicly available (*PhET Interactive Simulations, University of Colorado, http://phet.colorado.edu*).

Some of these virtual experiments can be done as extensions to the class discussions; in other words, these activities can be assigned as homework projects, thus enhancing the opportunities for inquiry and investigation they make available.

The simulations have been structured to allow the user to exercise both pace and variable control; studies have shown that simulations can often overwhelm students with low prior knowledge, due to quick and continuous changes that can overload working memory. Variable control can be particularly helpful in the development of exploration and hypotheses testing.

At the same time, the need for physicality is not neglected. Physics is and hopefully will always remain an experimental science; despite great advances in technologically rich environments, there is a basic need for physicality. The role of kinesthetic tasks is an area of considerable interest due to the findings concerning student retention and understanding of the material. To this extent there are nearly 20 experimental tasks included that require physical manipulation of variables.

I have endeavored to demonstrate that the approach taken in this book will benefit all readers, particularly those among you that tend to be intimidated by scientific concepts. I don't know what the readers' experiences have been, but mine have consistently shown me that there are many more students from the sciences who are interested in the arts and the humanities, than it is the other way around. I sincerely hope that with this book I can help change that!

I believe instructors cannot afford to neglect their responsibility to the audience (students); there are simply too many great and interesting aspects of waves that all students should be allowed to understand, since they will enrich their understanding of their preferred areas of study.

Instructors can decide on how to cover the material given their individual circumstances. The book has been primarily, although not exclusively designed for non-science majors, and students must possess some algebraic proficiency. If the instructor finds that students struggle with quantitative information, my recommendation is to concentrate on those chapters that don't require a significant amount of mathematical detail. However, don't neglect the value of exploratory tasks found in such chapters, since they may be more palatable and instructive for those students without requiring mathematical expediency.

There is no particular sequence needed to expose students to the many interesting aspects and applications of waves. Therefore, if some chapters must be omitted to facilitate student comprehension that may be hindered by a lack of algebraic proficiency, this would not constitute an obstacle for students to develop a basic understanding of wave motion.

Based on my experience with a class of non-science majors that needed a science course to fulfill a general education or distribution requirement, I decided to cover only the first six chapters.

When teaching the properties of light or those of sound separately, one can find applications in all chapters, and so it is a matter of choosing the relevant aspects to the topics that are found throughout the text. Given this scenario, it is quite feasible to cover at least the first nine chapters in a given semester.

Instructors can also choose chapters that they consider appropriate for their particular student audience. The first nine chapters are undoubtedly driven by content-specific properties of waves, while the last three are concerned mainly with a variety of applications that utilize many of these concepts and properties. Therefore, instructors can provide students with an overview of all the topics, by concentrating on the exploratory tasks exclusively. The narrative sections that often precede these tasks can be assigned as part of the background knowledge to successfully carry out the tasks.

Figures 2.1, 2.3, 3.4, 3.9–11, 4.1, 4.4, 4.6–11, 6.1, 7.8, 8.3, and 9.1 were constructed using Physical Science Images & Art (Qwizdom Inc.) used with permission.

Credits: Fig. 10.2-credit: Wikimedia Commons, Fig. 10.10 and that of the Exploratory Task on p. 196-courtesy of Imgur.

Long Island, NY, USA   Fernando Espinoza

Contents

The original version of this book was revised. An erratum to this book can be found at
http://dx.doi.org/10.1007/978-3-319-45758-1_13

About the Author

Fernando Espinoza is a professor with a joint appointment in the School of Education and the Department of Chemistry and Physics at the State University of New York (SUNY) College at Old Westbury, as well as an appointment in the Department of Physics and Astronomy at Hofstra University. He has over 25 years of teaching experience at the high school and college levels teaching astronomy, physics, Earth science, physical science, and in the pedagogical preparation of science teachers.

He has an active research agenda that includes numerous peer-reviewed publications, a textbook *The Nature of Science*, and a significant number of conference presentations. He serves as a reviewer for several journals, most recently as a member of the New York State Education Department's Science Content Advisory Committee, charged with providing feedback on the adoption of the common core science curriculum as part of the Next Generation Science Standards (NGSS).

Chapter 1
Introduction to Wave Phenomena

Why Is the Study of Waves Important?

Most information human beings are exposed to in our interaction with the world is in the form of waves. Our senses convey to us an enormous amount of information about the natural world, both externally and internally, that is predominantly processed as properties of waves. From sights and sounds to pressure variations involved in touch, as well as olfactory and taste sensations that exhibit patterns of change characteristic of alternating conditions.

In addition, many occurrences and events in a wide variety of experiences that are presented to us as information can be categorized as cycles or recurring instances of properties that can be understood in terms of those of waves. An understanding of wave motion can help us to describe phenomena that apparently don't have anything in common, in a way that enhances and promotes general knowledge.

To begin to understand waves, we need to realize that the condition of most physical systems that use the energy available for action as work is a state of equilibrium (static or at rest); however, dynamic equilibrium (involving motion) can also be seen in the context of variation as long as there is a balance in the changes, such as the relationship between the job market and unemployment. Conservation (something remaining constant) and symmetry (something remaining identical) are other properties of systems, provided these are closed ones. In such cases something remains unchanged, while repetitive changes can take place with reference to that condition.

The transfer of information, in general, rests on an understanding of the concept of a *signal*, where its properties are better aligned with those of a traveling wave than with those of a moving object. Consider our responses to signals representing electrical impulses that generate the many body sensations we instinctively recognize and react to in various ways. The strength of those signals changes according to a property of waves, the *amplitude*; it determines how much energy is transmitted. It would be very odd to regard changes such as the intensification or diminishing of those signals in terms of the properties of particles. Nevertheless, in the very strange world of subatomic or quantum phenomena, that appears to be the case, although

© Springer International Publishing Switzerland 2017   1
F. Espinoza, *Wave Motion as Inquiry*, DOI 10.1007/978-3-319-45758-1_1

whether such properties are associated with particles or waves depends on the measurement. Consequently, at the level of perception the strength of signals is predominantly understood in terms of wave properties.

Another important concept that shares features with those of waves but not particles is that of a *field*; an example of such a property would be that of being infinitely extended. A field can be effectively used to describe something that varies from place to place. Consequently, using concepts derived from wave phenomena can help us to understand properties of nature that range from the infinitesimally small to the largest scales in the observable universe. Consider one of the most bizarre ideas in modern physics, the concept of entanglement, where a subatomic particle can communicate with another instantaneously even if separated by enormous distances. This would be extremely difficult to understand, even conceptualize in ways other than using the properties of ideas such as fields.

A particular source of difficulty in understanding many properties of wave phenomena, even at the everyday level of experience has to do with their speed of transmission through space or through various material substances. Human reaction times are categorized according to how long it takes for us to respond to various stimuli. For example, it is known that the average response time for visual stimuli is about 0.25 s, for audio stimuli it is about 0.17 s, and for touch it is about 0.15 s.

Conceptual Task

Consider the role that human reaction time plays in driving a motor vehicle. Generally speaking, this activity that has become necessary for many people requires a high level of attention to a number of processes and events, with some of them being unpredictable. It is not a good idea to "tailgate" or to drive too close behind other vehicles, especially if one is traveling at fairly high speeds. Suppose you are driving at approximately 35 mi/h on a road with some traffic, and the driver ahead of you moving on another lane in the same direction suddenly moves onto your lane.

We assume that you will have space to maneuver and this example of course entails some other assumptions, among them the speeds of the vehicles in question. We shall assume that they are constant; If the other driver is traveling with higher speed than your own, you will have more time to react and possibly avoid a collision. This is a result of the difference in speeds where the other vehicle will cover a longer distance than yours. The situation also entails that the only time available is 0.25 s, the average human reaction time. Use the formula Distance $(D) = $ (speed) (time). We also need to use some conversion factors; 1 mile ≈ 5240 ft, 1 h $= 3600$ s.

(A) If the other vehicle is moving with the same speed as your own (it should be obvious that if the other vehicle's speed is lower than yours, the situation is more critical still), what should be the shortest distance in feet between the vehicles to avoid a collision?

(continued)

(B) Suppose now that you are traveling at 65 miles/h on a highway flowing with the traffic and the vehicle ahead of you on the same lane is 20 ft away. Would you have enough time to react if the other vehicle suddenly came to a stop?

(C) Texting while driving is something one should never do! Suppose you receive a text and your phone is on the passenger seat; if it takes you about 1 s to look at it, how far will your vehicle travel in each of the above cases?

At the same time the calculated speeds for muscle movements vary from the fastest signals of 268 miles/h or about 120 m/s, to those for touch of about 80 m/s, and finally to the slowest ones for pain sensation of about 0.60 m/s.

Exercise
Convert the numbers 120 m/s and 0.60 m/s into feet/s (1 m = 3.3 ft).

Using the figures above, along with the relationship between distance, speed, and time given by $d = v\,t$, one can estimate the reaction time for the fastest muscle signals and what this means in terms of mental awareness of certain body movements.

Assume the distance traveled to the brain is roughly the length of an arm and the neck to be about 1 m, the time for signals to be recorded is $t = d/v \rightarrow t = 1\text{m}/119\text{m}/\text{s} = 0.0084s$ which is roughly 8.5 ms.

Now comparing this time to the reaction time for the sense of touch (150 ms), we can see that it takes a lot longer to feel touch than to be aware of one's own arms. To experience this, close your eyes and wave your arms; you undoubtedly know where your arms are at all times, since the sensation is almost instantaneous. We can also see why it takes some time to react to a painful sensation, since those are the slowest signals. The anatomical features of our sensory organs can also be understood in terms of the range of sensation and perception, as functions of the frequencies (a property to be formally introduced in the next chapter) of those waves we are exposed to, particularly in the case of vision and sound.

At the heart of wave phenomena lies the concept of a pattern. Patterns of change can be steady or *trends*, or changes that occur in *cycles*, and changes that are irregular or *chaotic*; sometimes a system may exhibit all three types. For instance, the daily weather isn't always predictable, but the climate of a region often is. Individual human behavior in isolation may be unpredictable, but we can become predictable when acting as parts of large groups.

Predictability is one of the most useful properties of behavior described as wave phenomena, since the perceived patterns of repetition can enable us to determine future conditions, based on those previously or currently observed. Consider the examples shown in Figs. 1.1 and 1.2 of data collected that can be represented as wave phenomena, despite initially not appearing to display such behavior.

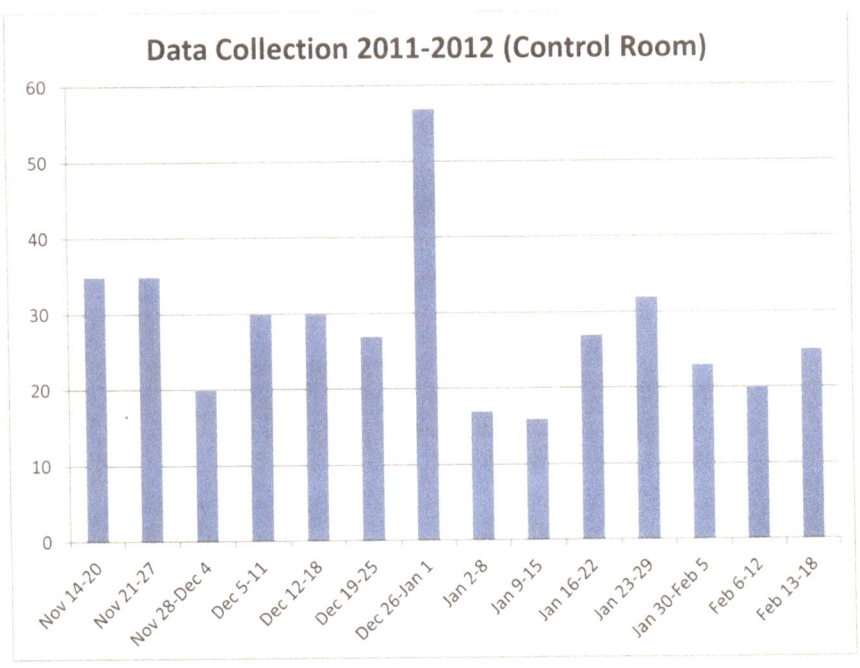

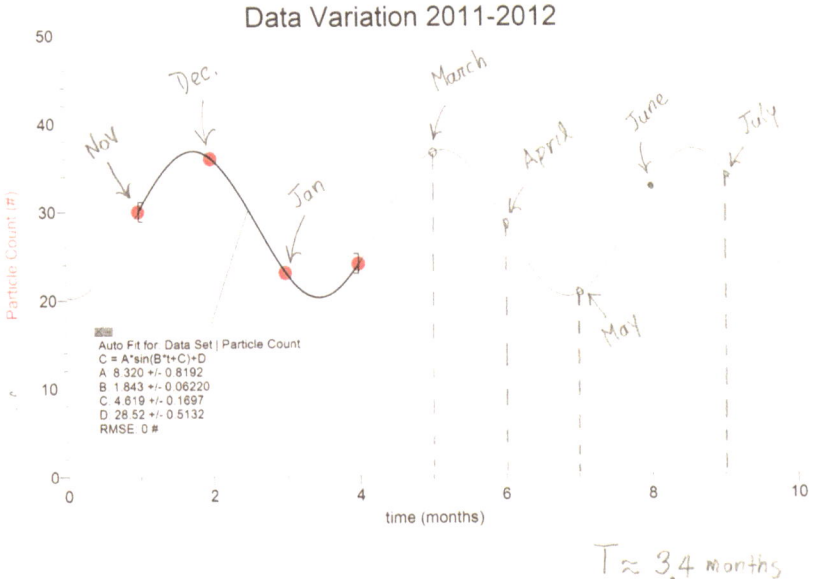

Fig. 1.1 Measurements of indoor air pollution in terms of particulate matter (microscopic dust particles) can be analyzed as waves where the amounts change throughout the year. The pattern shows repeated values that can be used to draw conclusions and make predictions about the data

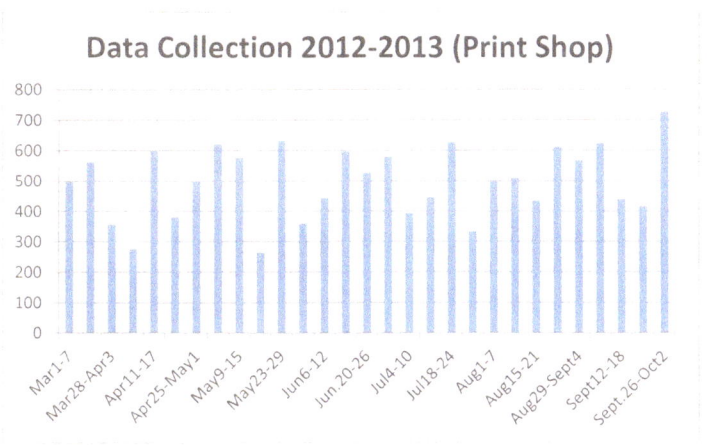

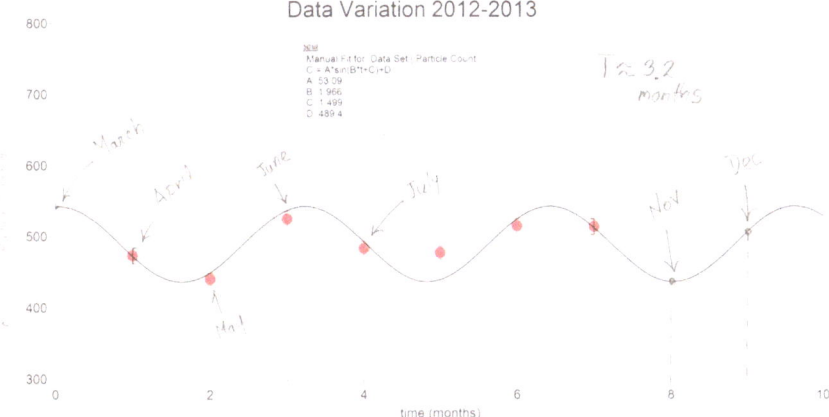

Fig. 1.1 (continued)

As seen in these examples data collected as numbers or other quantities can be represented as waves. We are daily exposed to information about many aspects of life that may not initially appear to form trends; however, if there is a way to display such information as waves, it could become much more significant by exhibiting features like: (a) repeated changes between a maximum and a minimum, (b) time when values repeat, (c) length or duration of changes, and (d) past and future changes.

Conceptual Challenge
When we use the terms "crime wave" or "heat wave," what do we mean to express?

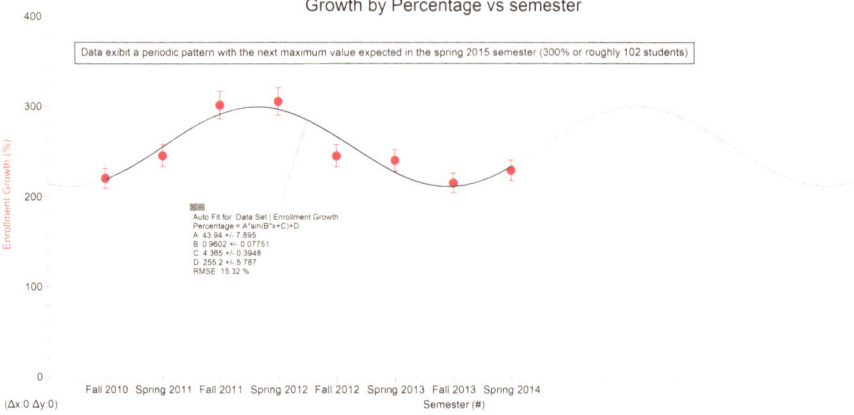

Fig. 1.2 Number of applications to a graduate program expressed as a percentage based on the first group of applicants. The *pattern* shows that the applications appear to change in a cycle and predictions can be made about likely future changes

There are of course many other examples of phenomena that can be represented as exhibiting wave-like behavior. Among these are biorhythms, motions produced by sports fans in stadiums, weather patterns, earthquake waves, and other types of natural disasters like a tsunami. There are other well-known phenomena that are increasingly being understood and explained as consequences of wave properties; examples of these are photosynthesis by plants, bird navigation, and chemical exchanges such as those involved in the perception of smell.

The essence of all waves is a vibration or oscillation; however, a significant consideration in our study of wave phenomena is the fact that whenever we speak of waves we are referring to the motions of individual particles or material objects, whose vibrations produce such waves. Therefore, one might ask if waves really exist as separate entities or as constructs. A construct is essentially something not directly observable, but a mental product based on properties exhibited by observable objects. An example is the concept of density, which is expressed as the ratio of the amount of matter contained in an object, and the volume of space it occupies; the latter two are experimentally determined but the former one, the density isn't directly measurable but instead defined in terms of these.

Conceptual Task
The electron is usually taken at the level of perception to be the fundamental unit of charge, and it can also be considered a human construct. It was discovered in 1897, although it was possible to describe the electrical properties of matter before its discovery using other concepts. Did electrons exist before 1897?

Wave propagation in a material or medium depends on the medium's response to a disturbance; the propagation characteristics depend only on the medium, and not on the nature of the disturbance, contrary to what many students believe based on everyday experiences. This example illustrates the approach to be followed in this text; it is imperative that we use the familiarity of concepts associated with waves so that students can use them to make sense of the many abstract properties that waves have and that are often difficult to comprehend. To this end, experimental tasks are designed to be exploratory rather than confirmatory.

The emphasis is on inquiry as the didactic approach to the presentation of the material. According to learning theory there is an inverse relationship between the degree of abstraction and retention of the material. In other words, material presented in the traditional way that textbooks have, with text, pictures, and other more recent ancillary methods that emphasize generalized phenomena, is retained very little by the learner. Information presented as dramatized/contrived, and purposeful (real life) situations based on learners' prior experiences on the other hand, results in the greatest amount of retention for individuals. It also effectively addresses misconceptions, which are often impediments to learning new material.

Theoretical Background

In order to properly understand waves a number of terms need to be introduced since these constitute the terminology necessary to describe all waves, and their precise use allows one to effectively apply them in the many situations and tasks where such understanding informs our knowledge of nature.

As will be formally introduced in Chap. 2 many of these terms have a quantitative definition or representation, besides a qualitative aspect that facilitates the ways in which we can describe them. Our objective is to be both accurate and precise in our treatment of wave phenomena. Accuracy can be defined as the degree or measure of agreement between the description (both qualitative and quantitative), and the properties being described. Precision is defined as the consistency with which we use such descriptions. Let's use an example to illustrate both terms.

Suppose we wanted to hit a target by throwing darts as a group, and we could control the most important variables involved in successfully repeating the task, such as the distance from the target, the height, and the force of the throw. Figure 1.3 represents two outcomes of the activity.

Fig. 1.3 Outcomes of the dart task; the crosshair represents the target, and the dots individual shots. According to the above definitions, which outcome is more accurate, and which is more precise?

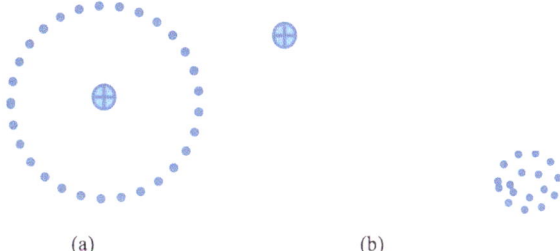

(a)   (b)

Both accuracy and precision are extremely important in all scientific measurements, as well as in other areas where events or processes occur, and descriptions need to be provided. Consider broadcasting and journalism where reports about newsworthy events (the crosshair) are provided by different agents (the dots), the need for accuracy and precision is no less important here than in scientific work, don't you think?

As stated before, a displacement or disturbance is necessary to create a vibration about a reference or equilibrium point, the material or medium in which the vibration takes place must exhibit the property of restoring the initial undisturbed condition after a period of time. When the movement is repeated (away from equilibrium and returning to it) an oscillation results. Whenever the oscillation takes place during a fixed amount of time, it will be called periodic (a term that will be explained in detail in the following chapter). A special case of periodic motion is described by Hooke's Law, discovered by Robert Hooke in the seventeenth century. It occurs in situations where the force acting on an object is proportional to the position of the object relative to some equilibrium position (like the motion of a mass attached to a spring or a pendulum). The proportion or relationship between the force and the displacement requires that the force be always directed toward the equilibrium position, in which case the motion is called **simple harmonic motion**, and that the quantitative variation between them be linear. This way the graph representing the motion looks like a straight line.

A fundamental idea in the study of all wave phenomena, Hooke's Law can be introduced in either of two ways: (1) horizontally—as a mass m attached to a spring, the mass being free to move (provided it rests on a frictionless surface), and (2) vertically—as a mass m suspended from a spring, as illustrated in Fig. 1.4. In both

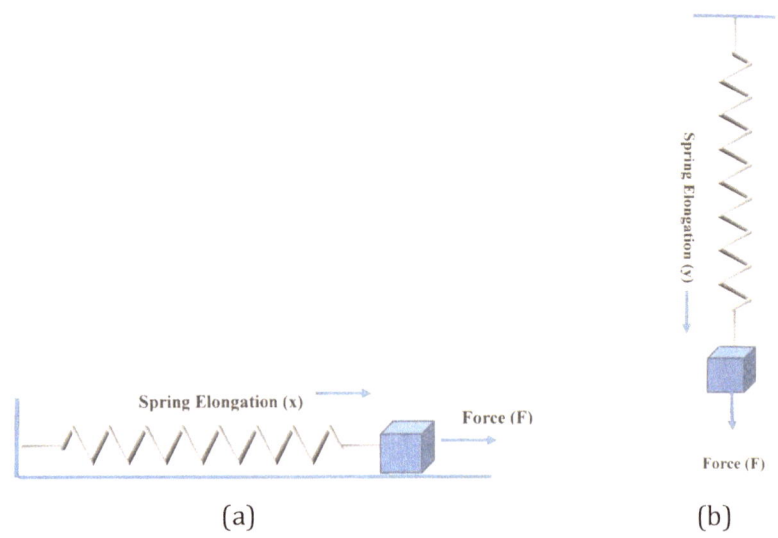

(a)   (b)

Fig. 1.4 Horizontal (**a**) and vertical (**b**) representations of a spring being elongated due to a force on the attached mass. Either representation can be used to introduce Hooke's Law, although (**a**) is typically chosen with the proviso that the surface on which the mass lies is frictionless

situations the spring could initially be un-stretched, in (a) the mass could still be attached, but the mass shown suspended from the spring in (b) couldn't be there. We could still show equilibrium in (b), such that the force that causes the elongation (gravity) can be incorporated into another expression (gravity minus the spring force) that also represents the spring being at rest. Nevertheless, in both cases the relationship expressed as Hooke's Law can be represented by the spring experiencing a force that according to Newton's third law of motion (action–reaction) leads to the spring reacting to it by moving back to equilibrium.

Choosing (a), when the spring is neither stretched nor compressed, the mass-spring system is at the *equilibrium position*, meaning that $x=0$. Such a system will oscillate back and forth if disturbed from its equilibrium position.

When stretched or compressed by a force of magnitude F, Hooke's Law states that the spring will react with a force $F_s = -k\,x$. F_s is the magnitude of the restoring (spring) force, being always directed toward the equilibrium position. Therefore, it is always opposite to the displacement from equilibrium, hence the negative sign in the equation; k is the spring constant and x is the displacement.

Clarification

You may have noticed the use of terms that you might not be familiar with, such as "magnitude" and "displacement" in the last paragraph. There is a term in physics that is useful in describing quantities representing properties of nature that need two things to be properly defined. It is called a *vector*, and it requires a magnitude (a number with units or dimensions), as well as a direction. An example would be the motion of a football player that upon catching the ball and running into an opposing player is pushed back before stopping. The total distance the player covered would not take into account which way he moved; however, the gain would need to include both distance and direction, since it is a vector (in this case it is the displacement). One can see that in terms of the game there is more information contained in knowing the gain, as opposed to simply knowing how much distance he covered. Another example would be an airline pilot on approach being given information by the control tower; it is more useful to know what distance, as well as what heading or direction the plane is on, than just knowing how far away it is.

Forces are examples of vectors, hence the need to specify the amount of force (the magnitude) as well as its direction in the case of Hooke's Law.

It is extremely important to distinguish between F, the force that causes the initial disturbance, and F_s the elastic or restoring force, since the latter is the one that describes the *behavior* of the spring.

There is an interesting historical context in which Hooke made his discovery. Anagrams (a type of riddle to publicly claim priority in a discovery, while preventing anyone else from knowing what it was) were popular in Hooke's time. As Galileo had done some 60 years earlier when he discovered the rings of Saturn (actually he couldn't make out the rings with his telescope, but cleverly used an anagram describing them as giant ears around the planet), Hooke published his anagram in 1676 as **ceiiinossssttuv**. The arrangement of the letters in Latin was revealed by him 2 years later as "ut tensio sic vis" which translates into English as "as the extension, so the force" [1] which has come to be known as Hooke's Law, in symbolic form $F = -kx$.

Exploratory Task
Determining the accuracy and precision of predictions and measurements with a set of springs.

Investigating the properties of springs allows one to apply Hooke's Law to obtain the values of unknown masses from the graphical relationship established between the forces of known masses and the elongations of various springs. An online simulation can be used (http://phet.colorado.edu/index. php).

Choose the *mass-spring lab* from the available choices, make sure the screen looks just like the figure below.

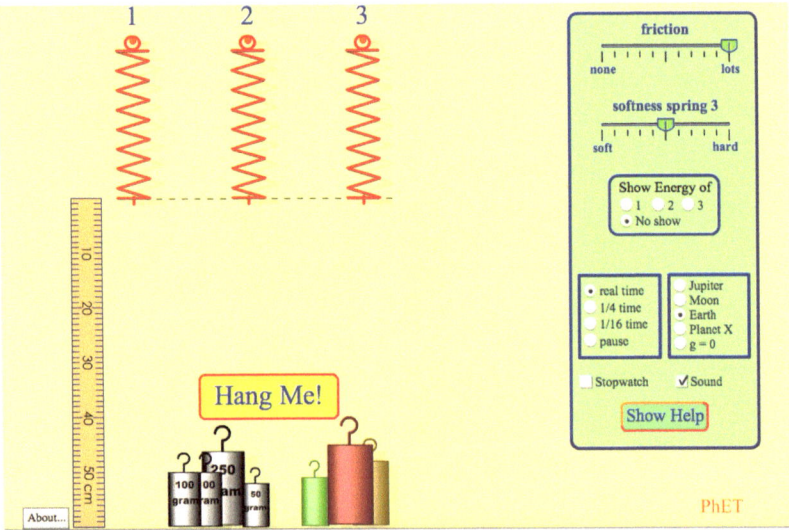

(I) The springs are identical (they have the same value of k), and they stretch different amounts from the equilibrium position (the dashed line) due to the different masses.

• Note the vertical displacements on the springs (after you attach the masses the springs pull back, but *they don't return to the equilibrium position*). Record the displacements in the table that follows. Make sure to divide the

(continued)

value by 100 to convert the cm to meters. For each mass, convert it to Kg by dividing the mass in grams by 1000, and then multiply the value by 9.8 m/sec^2 to obtain the force acting on the mass. Record each force on the same table. Then remove the known masses and replace each one with its equivalent size (the 50 g one with the smallest one, the 100 g with the middle one, and the 250 g with the largest one).

- Record each displacement of these unknown masses ( ), ( ), ( ).
- Based on the vertical displacement, and comparing it to that with the closest known mass, can you predict the value of each unknown mass?

M_1 (the smallest) _____
M_2 (the medium sized) _____
M_3 (the largest) _____

(II) Now plot the data obtained from each of the known masses and the vertical elongation (the displacement) of the particular spring it is attached to. The force (the product of each mass in kg and the local gravitational constant, $g = 9.8$ m/s^2) along the vertical axis and the displacement along the horizontal one.

Object	Mass (Kg)	Force (mass x 9.8 m/sec^2)	Displacement (meters)
M_1			
M_2			
M_3			

Graph of Force vs Displacement

Draw the "best fit" line (a line that connects the dots) and determine its *slope*.

$$\frac{\text{Difference between the larger and the smaller of any two of the three forces}}{\text{Difference between the larger and the smaller of any two of the three displacements}}$$

This represents k the spring constant.

- Locate each displacement from those recorded in part (I) for the unknown masses on the graph on the horizontal axis; draw a line from each point to the best fit line, and then from the best fit line to the vertical axis to determine the value of the force for each point.
- Finally divide each value of the force by 9.8 m/s² to determine the mass corresponding to each point. This will yield the corresponding mass in Kg, which needs to be multiplied by 1000 to convert it back to grams. How do they compare to your predicted values in part (I)?

(III) Reflection

The springs in this simulation can be used over and over and they will not change since this is a virtual setting; however, in a real setting actual springs will change with repeated use.

a) How will such use affect the springs?
b) What property of the springs used in this simulation will show the change, and in what way?

Exercises

Using the same simulation as in the task above.

1. What happens to spring 3 when you change its softness to "soft" and then to "hard,", and then attach each of the masses to it?
2. Using the softness of spring 3 back in the middle of the range, predict what will happen to it when the 50, 100, and 250 g masses are attached to it, if you choose to do the experiment on the Moon.

   Prediction _____
   _____

   Now choose "Moon" from the choices in the lower right-hand side of the simulation and test your prediction. How did you do?
3. Change the friction rider to the middle of the scale (and back on Earth), and describe what happens to all the springs as each mass is attached to it.

There are many applications to other phenomena that are based on the consideration of what happens to a spring when it moves repeatedly back and forth after a mass has been attached to it. They are examples of the usefulness of Hooke's Law in describing the behavior of a spring, or an object moving in a similar manner (describing simple harmonic motion). The mathematical description of its motion forms the basis for the treatment of periodic motion (the back-and-forth or up-and-down motion repeats and oscillations result).

Let's use two examples to illustrate the usefulness of modeling situations or interactions that are essentially invisible, where the details are beyond the level of perception.

(1) We can visualize the concept of a field as a collection of points in space where an object experiences an effect (depending on the type of field) due to the action resulting from imagined springs that stretch or compress depending on where the object is. Figure 1.5 illustrates such a model.

(2) We can also visualize the microscopic behavior of matter by imagining the molecules that constitute material objects as being attached to each other with springs. Of course other models of matter have been historically used, and are currently modified such as the planetary one. However, for purposes of interactions and behavior under different conditions such as phases (solid, liquid, and gaseous), the overall dependence of states of matter on temperature can be effectively understood with a spring model. Figure 1.6 illustrates how this can be accomplished.

To summarize this chapter we discuss the ways in which the speed of the two waves this text concerns itself with has been determined. They are based on the same relationship that we already introduced in the discussion of the propagation of

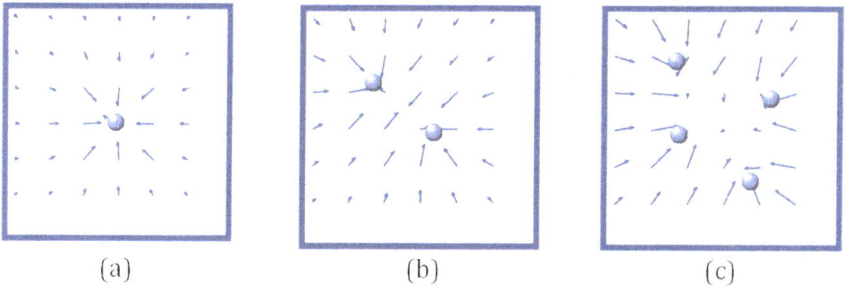

(a)   (b)   (c)

Fig. 1.5 Representation of a field and its effect on an object as a region of space filled with springs that stretch (*the length of the arrows*) depending on where the object is. The presence of a single object is represented in (**a**), that of two objects in (**b**), and that of several in (**c**)

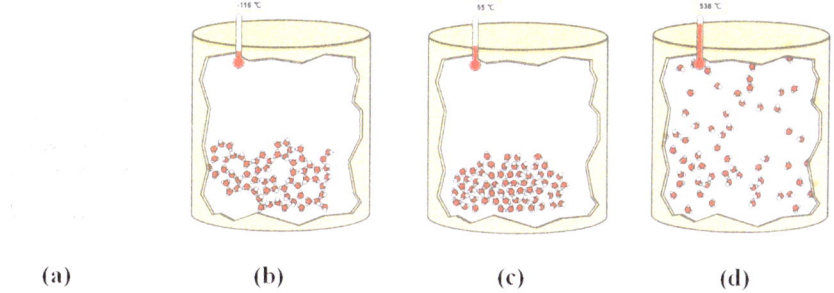

(a)   (b)   (c)   (d)

Fig. 1.6 Representation of a model of molecular structure using as a basis the cube in (**a**) that contains a molecule or an atom at each corner attached to other molecules or atoms with springs. In (**b**) the substance is meant to be a solid (ice) at −116 °C, in (**c**) it is meant to be a liquid (water) at 55 °C, and in (**d**) it is meant to be a gas (steam) at 536 °C. Using the spring model, what do you notice about the arrangement of the molecules that is different in each case?

signals through the human body. The relationship is expressed by the ratio of the given distance traveled by the waves and the time taken. Interestingly, both attempts to determine the speed of light and of sound initially took place in the seventeenth century [2].

In the case of light, the earliest known attempt was that of Galileo Galilei, who attempted to measure the distance between two locations (the top of mountains), and the time that it took for two lanterns to be covered during the night. As he found out, the time taken for light to travel the distance between the mountains could not be measured, as the covering of the lanterns seemed instantaneous. It became apparent that much longer distances were needed given such believed large speed for light. The first determination was provided later in the century by the Danish astronomer Olaf Roemer who observed the eclipses of one of Jupiter's moons (incidentally discovered by Galileo). Roemer determined a discrepancy in the time between the eclipses, increasing when the Earth was moving away from Jupiter and decreasing when the Earth was approaching. He correctly surmised that if the speed of light was infinitely fast, there should be no difference between the measured times for the eclipses' duration.

In Fig. 1.7 the lower left-hand side shows the orbit of the earth around the sun; the larger circle represents the orbit of Jupiter with the small circle being the orbit of one of Jupiter's moons, Io. In (a) when the moon has just gotten behind the shadow of Jupiter, its eclipse begins. The time Δt_1 that it takes the moon to emerge from Jupiter's shadow is measured when the Earth is at a distance D_1 from Jupiter (J_1). In (b) the Earth has now moved on its orbit around the sun to D_2, and the time Δt_2 for the eclipse has now increased. Roemer reasoned that if light had an infinite speed, there should be no time difference between an eclipse observed at (a) and at (b). In other words, it wouldn't have mattered where the earth was in its orbit in either case, Δt_1 should be equal to Δt_2. But they aren't.

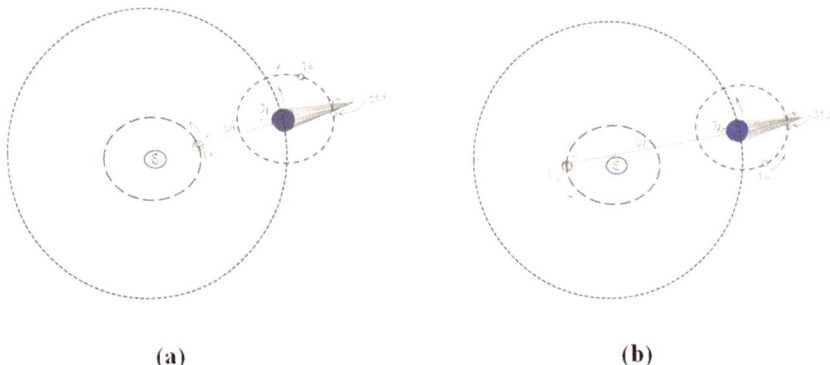

(a)   (b)

Fig. 1.7 The figure shows the time difference in the measurement of the length of the eclipse of one of Jupiter's moons, Io. In (a) the Earth is closest to the planet, and in (b) the earth has now moved on its orbit around the sun. The diagram is not to scale

Roemer's value, the first for the speed of light, was off due to imprecisions in the time measurement and the fact that the Earth's distance from the sun wasn't accurately known. However, further improvements in the determination of both quantities have led to a very precise measurement of the speed of light. It is currently taken to be roughly 300,000 km/s or about 186,000 miles/s.

In the case of sound, Isaac Newton was the first to attempt to determine its speed in the same century. The approach involved knowing the same two quantities, the distance taken by sound to travel and be reflected by a wall as an echo, and the time taken. Newton's technique was further improved and with increasing knowledge of the properties of air as an elastic medium or substance, it has come to be determined based on air temperature. We use the following expression for the speed of sound, taking into account that it isn't a constant as that of light is taken to be.

$$V = \left(331 + 0.60 T°C\right) m/s$$

where at room temperature (22–25 °C), the speed is roughly 345 m/s.

Virtual Experiment

A determination of the speed of sound can be performed online with a simulation depicting a series of pulses from a loudspeaker. The distance each pulse travels and the time it takes to travel the distance can be measured.

Using the figure below, how would you determine the speed of the sound from the loudspeaker?

Go online at (http://phet.colorado.edu/index.php); select "sound" from the Physics simulations, make sure that your screen looks exactly like the figure below.

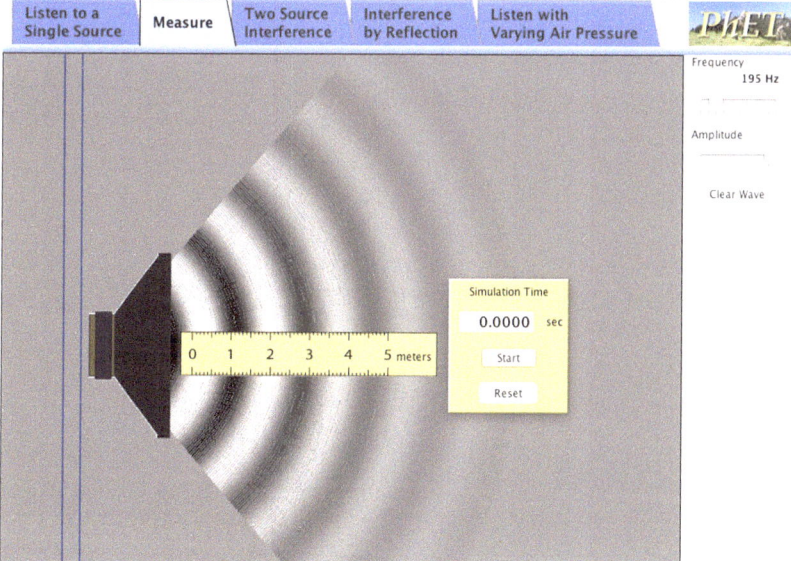

(continued)

Make sure to choose the Tab "Measure" and while the top rider on the right-hand side doesn't have to be the same number (195), the lower one should be at the right end as shown.

The task is to start the timer when the end (the left edge) of one of the dark bands passes through the zero (0) mark on the meter stick, and then to stop the timer when the same end passes the 5 m mark (if the band is too faint by the time it gets to 5, then choose 4 as the point to stop the timer).

Repeat five times and divide the distance you have chosen for the edge of the dark band to travel by the average time obtained from the table below. This will yield the speed of the sound produced by the loudspeaker.

Trial	Time
1	
2	
3	
4	
5	
Average	

While you cannot determine what the effect of temperature is on the speed, you may still compare your result to

$$V = (331 + 0.60\, T°C)m\,/\,s$$

Assuming room temperature (≈ 22 °C)

$V =$

What were the most challenging parts to deal with in performing the simulation?

We shall use V as the symbol for both *speed* and *velocity* in this book, disregarding the fact that one is a scalar, and the other a vector.

Experimental Task: Determining the Speed of Sound

There have been many attempts to measure the speed of sound, beginning in the seventeenth century and including efforts by Newton himself. He determined the speed by producing a noise that traveled along a long corridor and upon reflecting

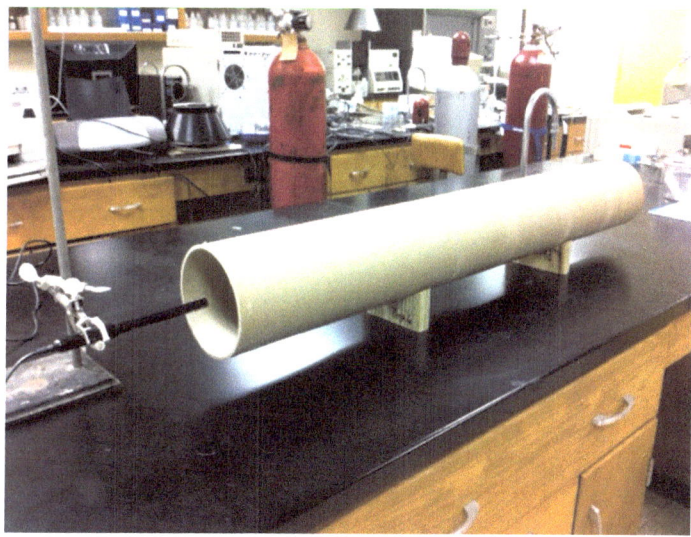

Fig. 1.8 Experimental setup for the determination of the speed of sound inside a tube that is closed at one end. The tube should be isolated to minimize vibrations that can contribute to the signal produced at the microphone

from a wall was heard as an echo. The timing mechanism he used was a pendulum, and the time taken for a trip of the pendulum was equated to the time taken by the sound to travel down the corridor and back [3].

Newton's result was inaccurate due to the timing mechanism, and the then unknown dependence of the speed of sound on the temperature of the air.

We are able to perform Newton's experiment in a modified way that allows for much more accurate determinations of the values involved in the speed of sound. The most significant one is the time measurement. By using a microphone one can measure the time taken by a pulse to travel the length of a tube closed at one end, as shown in Fig. 1.8.

This experiment version is a modification of the one developed by Vernier Software and Technology, and available in their *Physics with Vernier* manual of activities. The advantage of using it is that the file already has the settings for time measurement arranged to be displayed in the most user-friendly way.

Our objective is to measure both the length of the tube (the distance in the formula) and the time taken by the sound to travel this length, and then to use the formula

$$\text{Distance} = (\text{speed})(\text{time}) \text{ to find the speed of sound.}$$

As Fig. 1.8 shows the microphone needs to be held at the entrance to the tube; a sharp sound is produced, either by snapping one's fingers or clapping hands. Other objects may be used, as long as the sound produced is sharp. The microphone signal needs to be displayed by using an interface, such as Vernier's LabQuest. An example of the display is shown in Fig. 1.9.

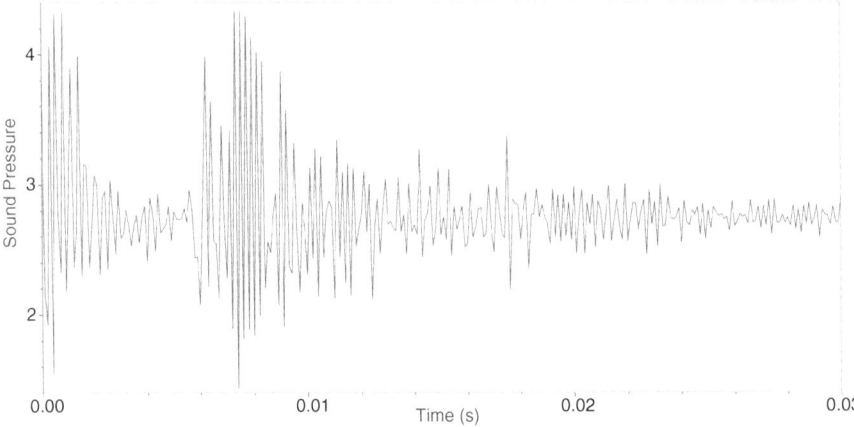

Fig. 1.9 Sample graph obtained by snapping one's finger at the position of the microphone; the initial peaks represent the signal produced, and the next set of large amplitudes represents the reflected signal picked up by the microphone

Table 1.1 Travel times for the signal generated by snapping one's fingers at the position where the microphone is located. The time recorded in each trial is that for the signal to travel the length of the tube and return to be picked up by the microphone

Trial	Total travel time (s)
1	
2	
3	
4	
5	
Average	

The signal generated is reflected at the other (closed) end of the tube and is shown as being also reflected inside other parts of the tube. The two largest sections, beginning with the leftmost one represent the sound being recorded by the microphone at the outset, and upon returning to it. The time difference between these large sections is consistent with what one would expect for sound to travel the length of the tube back and forth, to be picked up by the microphone.

Being that this time is the largest source of error in the experiment, it is advisable to record it several times to find an average value. The data are recorded in Table 1.1.

We must divide the average time from the table by two, as it is the time taken by sound to travel to the closed end of the tube, and back to the microphone.

We then measure the length of the tube and divide it by the time to get the speed of sound.

The accepted speed of sound at atmospheric pressure and 0 °C is 331.5 m/s. Determine the temperature of the room and use the following equation to calculate the expected value of the speed.

$$V = \left(331 + 0.60\,\text{T}^\circ\text{C}\right)\text{m}/\text{s}$$

Compare this value with the experimentally obtained value from the tube data, and determine the percent error.

Write a report including the following sections.

1. Objective
2. Brief procedure
3. Data, calculations, and results
4. Reflections-Analysis and discussion of sources of error.

Tasks on Accuracy and Precision

Experimental Task
Introduction to Measurement (Accuracy and Precision)
    The objectives are to introduce you to the concepts of accuracy, average, and precision, and to allow you to see how measurement plays such an important role in the determination of the values of those quantities considered *constant* in nature. Additionally a discussion of the sources of error inherent in every experiment that involves measurement will expose you to the realities of scientific work where all measurements involve uncertainty and the means to minimize it.
    Part I

- You will measure and record (Table I) the circumference and the diameter of several circular objects (see figure below).
- You will plot the data and determine the slope of the line; a determination of the class average (Table II) of this value will enable you, upon discussion, to understand the process involved in arriving at the accepted (standard) value of the relationship between the circumference and the diameter of any circular object.

Question: What do you think this relationship represents?

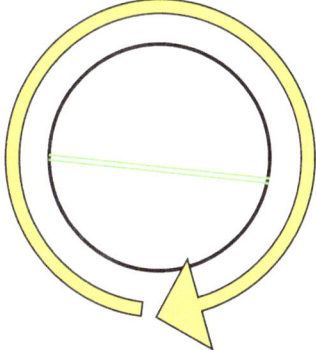

(1) Wrap the string around the circular object and determine its exact length (the circumference); then extend the string and measure its length with a ruler.
(2) Measure the diameter of the object with the ruler. Record the values in the table below.

(continued)

Table I Values of diameter and circumference

Object	Diameter (cm)	Circumference (cm)

Table II Values of the slopes and their uncertainties (uncertainty = [average - each value]. All uncertainties are written as positive values)

Slope	Uncertainty	Slope	Uncertainty
		Average	Precision

(Precision = average uncertainty) Accuracy: % Accuracy: % Precision:

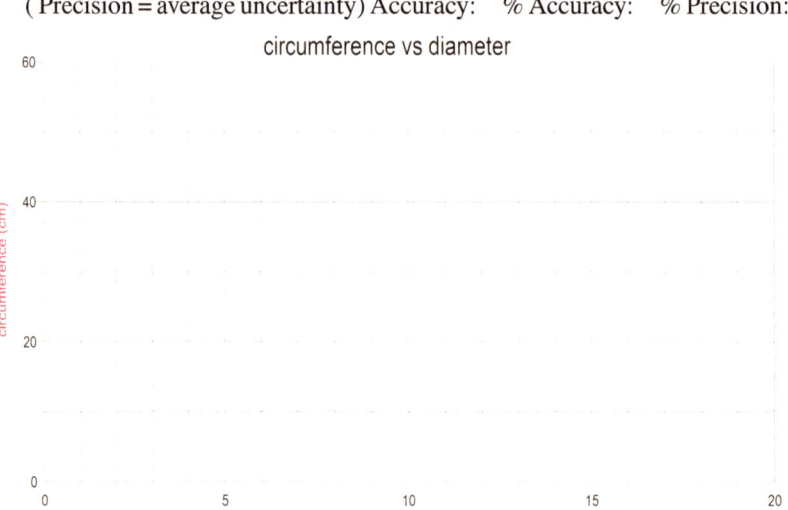

circumference vs diameter

(continued)

Reflections on the results (use accuracy, precision, and whatever you consider sources of error in the measurements as part of your comments):

Time (seconds)	Temperature (°) Analog Thermometer	Temperature (°) Digital Thermometer
0		
10		
20		
30		
40		
50		
60		
70		
80		
90		
100		
110		
120		

Part II. MEASUREMENT OF TEMPERATURE

Determine the temperature reading of the thermometer before placing in the palm of your hand and grasping it. As you hold it record the temperature every 10 s. Fill in the table below and then graph the results. *The time will be the independent variable, and the temperature the dependent variable.* Determine from the graph the maximum temperature (the point where the readings stabilize).

Question: Do you expect the relationship between hand temperature change and time to resemble the graph of circumference and diameter for a circular object?

(continued)

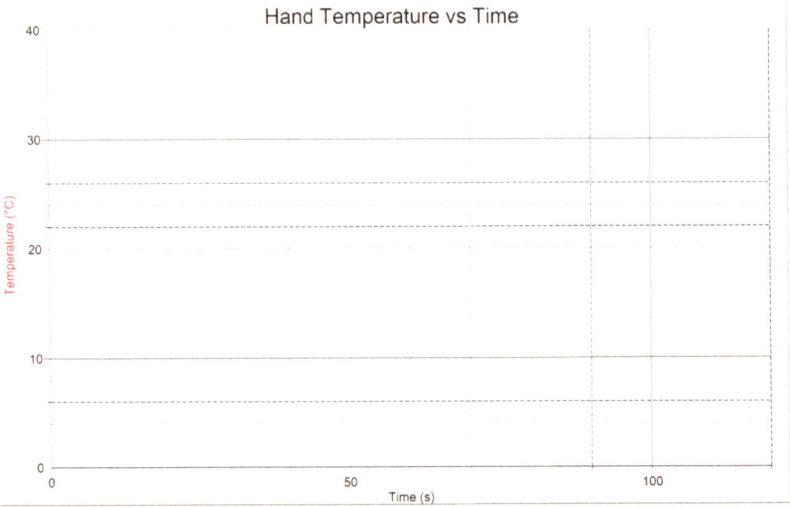

Maximum Temperatures =   (Analog)   (Digital)

Now fill in the table with the temperature maxima obtained by all members of the class. Note: if you only have access to either type of thermometer, the data can be collected and one of the columns is enough to display the relationship.

Analog		Digital	
Temperature (°)	Uncertainty (°)	Temperature (°)	Uncertainty (°)
Average	Precision	Average	Precision

(continued)

Analog results

Accuracy   % Accuracy   % Precision:

Digital results

Accuracy   % Accuracy   % Precision:

Part III. MEASUREMENT OF HEART RATE

Fill in the table with *your average* heart rate; determine the number of heart beats in 1 min, do it three times, and then average the result. If the values are fairly constant, three times is probably enough, otherwise do it a couple of more times.

Heart Rate (#)	Uncertainty (#)	Heart Rate (#)	Uncertainty (#)
Average Heart Rate		**Average Uncertainty (Precision)**	

Accuracy: **% Accuracy:**   **% Precision:**

Reflections on the results:

_____

_____

_____

_____

ACCURACY AND PRECISION IN WORKS OF ART

(Examples from Diego de Velazquez)

1-La Rendicion de Breda (The surrender of Breda).

(continued)

2-Don Juan de Pareja

(Wikimedia Commons)

Suppose the Hue (the most dominant color)—a property to distinguish one color from another is measured for several reproductions of the paintings above. The degree of consistency between the reproductions can be an indication of *precision,* whereas how faithfully the reproductions copy the original can be an indication of *accuracy.*

References

1. French, A. P. (1971). *Newtonian Mechanics*, Norton, p.227.
2. *'Asimov's Guide to Science,'* Isaac Asimov, Basic Books, Inc., (1972), pp. 342–347.
3. There are many reports and presentations about the historical background of Newton's measurement; among the best is a demonstration in the classic physics series "The mechanical Universe" available online.

Chapter 2
General Characteristics of Waves

As stated in the previous chapter, the advantage provided by an understanding of wave motion in describing apparently unrelated phenomena lies in the fact that all waves share the same properties and follow the same rules. While a detailed analysis of wave motion involving the mathematical description and representation of its properties requires a certain level of mathematical proficiency, many if not most of the characteristics of waves necessary for a thorough understanding of wave phenomena can be acquired without extensive mathematical manipulation.

Activities Designed to Elicit Prior Knowledge

Consider the following situations (in all likelihood you have an idea about each one, but all will be better understood once you become familiar with the properties of waves):

1. Go online at (http://phet.colorado.edu/index.php); select "sound" from the Physics simulations, make sure that your screen looks exactly like the figure below. The Tab selected is "Listen with Varying Air Pressure." Once you hit the Play button you will hear a sound, before you click on the box that says "Remove Air from Box" can you predict what is going to happen to the sound? Make your prediction and then follow the simulation.

(continued)

© Springer International Publishing Switzerland 2017
F. Espinoza, *Wave Motion as Inquiry*, DOI 10.1007/978-3-319-45758-1_2

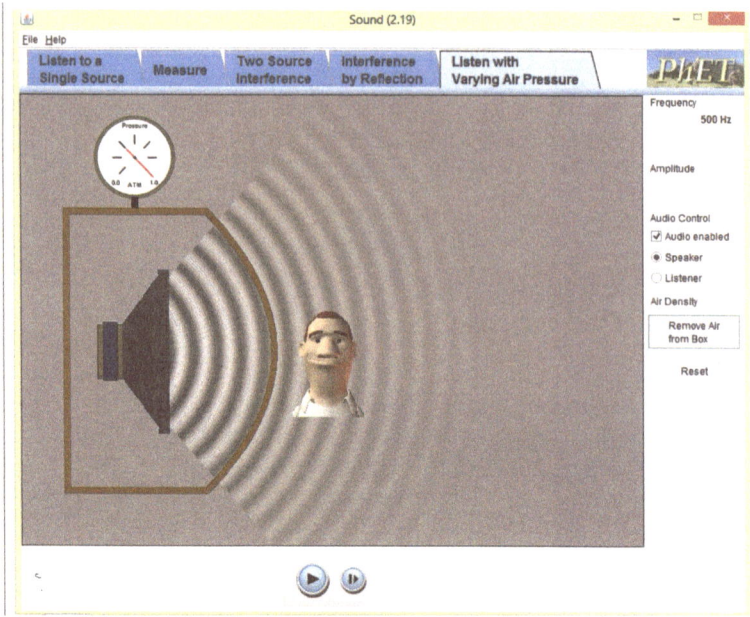

Describe what happens and provide an explanation for it; how does the
outcome compare to your prediction?

2. Suppose you and a friend are talking alongside a building as you move
away from each other and at some point one of you turns a corner of the
building as you continue to talk. Why can't you see each other but can still
hear one another?

3. If you look at the full moon as it appears above the horizon, does it look
different if you look at it standing and staring directly at it, and if you look
at it after you turn around, bend over and look through your legs?

4. Why do native Americans in some movies put their heads to the ground to
hear whether a stampede is coming their way?

5. Is the public sufficiently aware of the potential dangers of cell phone radiation? Why would anyone trying to make a phone call from a cell phone in
a rural area be exposed to more radiation than in an urban center?

6

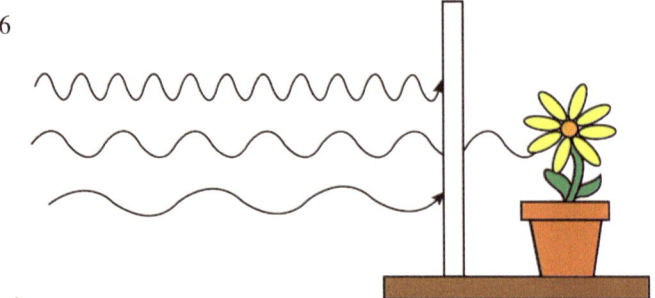

(continued)

The diagram on the left shows light striking the window of a room where there is a plant on the windowsill. According to the diagram, waves of different length strike the glass; what is happening in terms of the properties of these waves and how does that affect the plant?

The easiest way to approach the study of waves is to begin with their most basic component, a *pulse*. It is a single disturbance (like a push or a pull) to an object or a collection of objects (a system) that causes a movement away from its initial location, and to which it aims to return. This property of objects is best defined as a "springing" back to the initial condition, and the disturbance either dispersing or remaining intact as it propagates through the object or the system.

Exploratory Task

Does the speed of a pulse in a rope or a string depend on how much force is applied to them by shaking the hand more vigorously?

ANSWER:

What are your reasons for the answer?

Now test your answer by using a simulation at http://phet.colorado.edu/index.php.

Choose "Wave on a string" from the available choices. With the simulation open, make sure it looks like the diagram below.

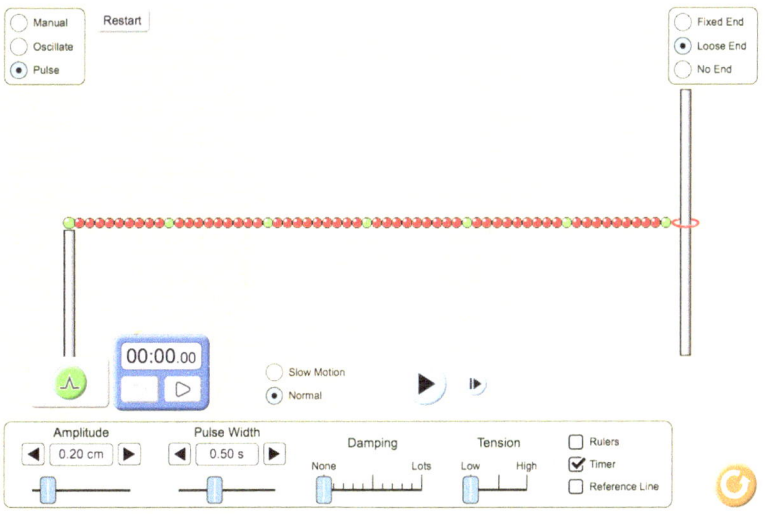

- Click *as simultaneously as possible* the Pulse and Play buttons and determine how long it takes for the disturbance (the pulse) to reach the ring at

(continued)

the end. Stop the timer and record the value. Repeat for "Amplitude" (the equivalent of the amount the hand is moved) values of 0.50 and 0.90.
- Now select "Amplitude" back to 0.20 and then change the tension to the middle of the scale; repeat the measurement of the time. Finally, change the tension to high and repeat once more, then fill in the table.

Amplitude	Tension	Timer Reading (seconds)
.20	low	
.50	low	
.90	low	
.20	middle	
.20	high	

What does the table suggest?

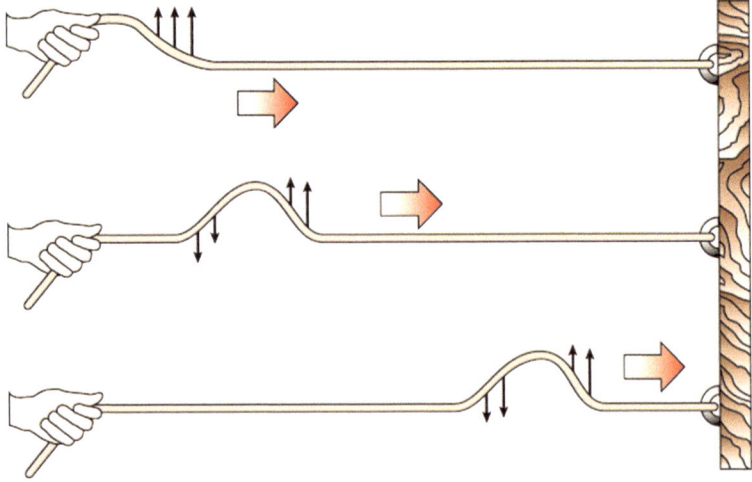

Fig. 2.1 Illustration of a pulse resulting from the motion of the hand that sends a disturbance along a rope attached to a wall at the other end. The *thin arrows* represent the motion of the rope components and the *thick arrows* represent the motion of the pulse. It is important to distinguish the motion of the rope resulting from the initial disturbance caused by the hand movement, from the direction of propagation of the pulse

Figure 2.1 shows a rope through which a pulse propagates; the ability of points on the rope to return to their initial position after the pulse has passed is a result of the property described by Hooke's Law.

There are many ways in which pulses can be created depending on the material or medium through which they propagate; for now let's concentrate on the properties of pulses as they move through a rope since these are common to other types.

A distinctive property of pulses is that they can go through one another as they interfere. The use of the term interference is different in this context from that of everyday language, where interference is used as an impediment or obstruction. When pulses interfere they do not collide or bounce off each other as objects do when they meet. Instead the various points displaced on the rope will reinforce or diminish, comparable to adding or subtracting each other depending on their orientation. This results from a property known as superposition. Additionally, when reaching a boundary between the rope and another object (such as at the point on the wall where the pulse in Fig. 2.1 ends), or between different types of ropes, the orientation and height of the pulse may or may not change, depending on the properties of the ropes (such as thickness, density) or the rigidity of the boundaries (whether fixed or flexible).

Exercise

Based on the previous section, predict what happens to the pulses when they meet at the origin by drawing the resulting shape of the rope at 0 for each of the figures that follow.

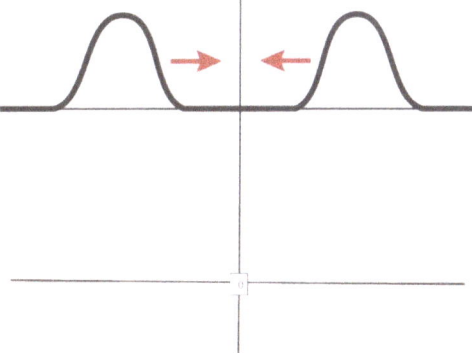

(continued)

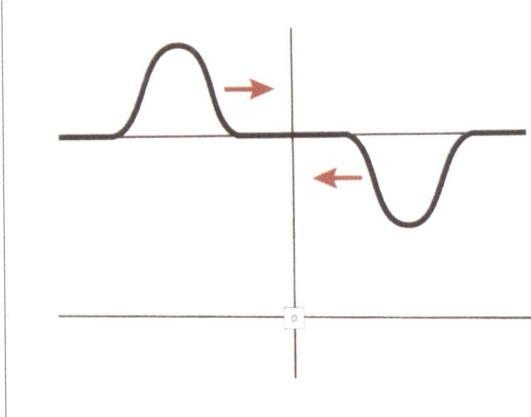

In order to begin our discussion of wave characteristics, it is useful to provide a simple example of someone attempting to transfer energy (in the form of heat and sound) to a wall located at a certain distance.

When energy or other forms of information are to be transmitted through space, this can be accomplished by either sending the energy as contained in an object, or as contained in a pulse, or as a wave. These three cases are shown in Fig. 2.2.

One of the defining characteristics of a wave is that unlike a particle, it cannot exist at a single position or location in space. The wave is instead spread out. In the example the bowling ball and the rope represent material objects containing many particles. However, the way they each transfer energy is very different.

In addition, particles carry matter from one place to another, but not waves. The motion of the matter that constitutes or makes up the above wave (the mass of the rope) will be in a direction other than the one the energy travels. The motion of the matter components of a wave can be along certain directions, but it isn't from where the wave begins to where it ends.

Waves come in many forms and shapes but they all have similar characteristics, which can help to develop a general understanding of their properties. Some examples of waves will be:

(1) A pebble that hits a still water surface, the resulting circular wave or disturbance spreads out in all directions from the point of impact. An object floating on the disturbed water will move both vertically and horizontally about its original position, but it is not displaced along the wave.

(2) A string that is plucked; the neighboring pieces on it pull on each other when a displacement is transmitted along the string. All particles of the string will move the same way as that caused by the initial disturbance, regardless of the wave's speed.

(3) A sound created by a falling object that hits the ground is also a wave that is made up of regions of compression and expansion of the air.

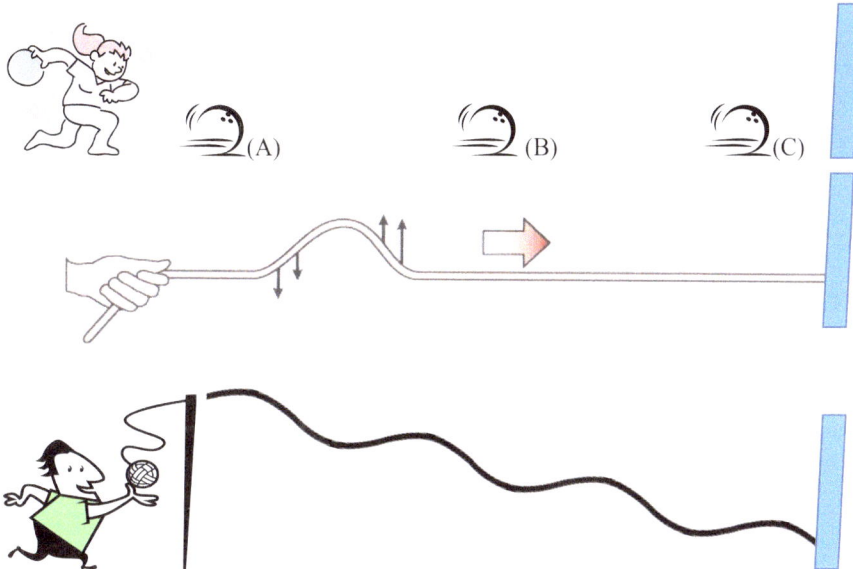

Fig. 2.2 *Energy Transfer to a Wall*—The *top part* shows a person throwing a ball against the wall on the right; the energy transmitted to the wall will be where the ball is at the three locations shown. The *middle part* shows a single pulse being created and sent through a rope to the wall; the energy in this case will be contained in the pulse, but it will be spread throughout the length of the rope. The *bottom part* shows that if the person instead shakes the rope several times and sends the energy to the wall through the rope this way, more than one pulse is generated, and this constitutes a wave

(4) The reception of television and-or cable programs that result from electromagnetic waves being broadcasted by specific providers or sources.

Classifications

There are two major ways to classify waves:

I.

(A) *Mechanical waves*—Some physical medium or material is disturbed. The wave is the propagation of such a disturbance through the medium or material. Examples are sound, water, earthquake waves, and waves in a rope or string.

(B) *Electromagnetic waves*—No physical medium or material is required for such waves to propagate, although space-time can be considered as a medium. The point is that electromagnetic waves can carry information without the need of a material substance, as compared to every other type of wave. Examples are light, radio waves, and X-rays.

Both examples are shown in Fig. 2.3.

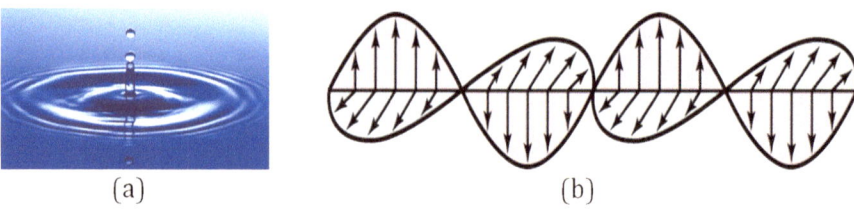

(a)   (b)

Fig. 2.3 (**a**) Mechanical wave—the ripples make up the wave that propagates in all directions. (**b**) Electromagnetic wave—the *arrows* represent the oscillations of the electric and magnetic fields

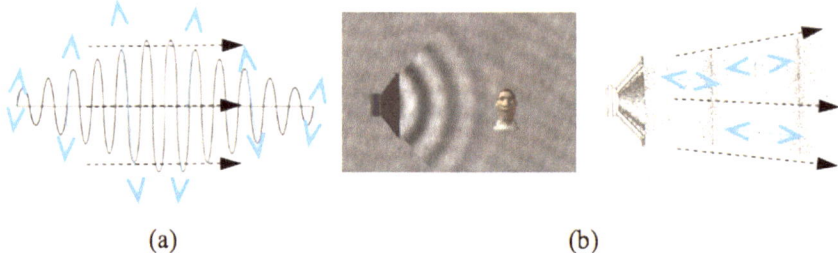

(a)   (b)

Fig. 2.4 (**a**) Transverse wave. (**b**) Longitudinal wave; the longitudinal requires further elaboration since the first part shows a loudspeaker emitting the sound while the second part shows the detailed motion of the air. In both (**a**) and (**b**) the *solid arrows* represent the motion of the components of the wave, and the *dashed ones* represent the direction of propagation of the wave

II.

(A) *Transverse waves*—Those where the particles or wave components oscillate at right angles or perpendicular to the way the wave travels. Examples are electromagnetic waves, secondary/shear (s-earthquake waves), and the wave shown in Fig. 2.2.

(B) *Longitudinal waves*—Those where the particles or wave components oscillate along the direction the wave travels. Examples are sound waves, primary/compressional (p-earthquake waves), and those produced by compressing or stretching a slinky.

Both examples are shown in Fig. 2.4.

Exercise

Imagine yourself in traffic along a highway and try to visualize how the motion of the vehicles can be represented by a wave. Since there are areas of congestion and expansion in terms of the space between vehicles as they move, it would be decidedly dangerous to drive in such a way that one of the two types of waves in Fig. 2.4 could be approximately represented by someone weaving in and out of lanes. However, the other type of wave invariably results as traffic flow varies. In what ways is the motion of the vehicles similar to that type of wave, and in what ways is it significantly different?

There are some types of waves that consist of a combination of transverse and longitudinal motion, such as surface water waves, where the motions of the water particles or wave components can be visualized as being circular, as seen in Fig. 2.5.

A generic drawing can be used to introduce the main properties of a wave. The diagrams are similar to facilitate the comparison between the determination of the wave's length and the time it takes to complete a cycle. Note that the units for the horizontal axes are different as the labels indicate, while those for the vertical axes are identical. The use of the term displacement is meant to emphasize that the motion of the particles that make up the wave is in a defined direction from the equilibrium position, represented by the horizontal dashed line. The range is from 0 to 10 up or along the positive direction, and down or along the negative one, as shown in Fig. 2.6.

Fig. 2.5 A complex wave that represents the familiar undulation of water waves; as one follows the sequence from 1 to 9 each point moves on a circle as indicated by the *small arrows*, but the collective motion is as indicated by the *dashed arrow*

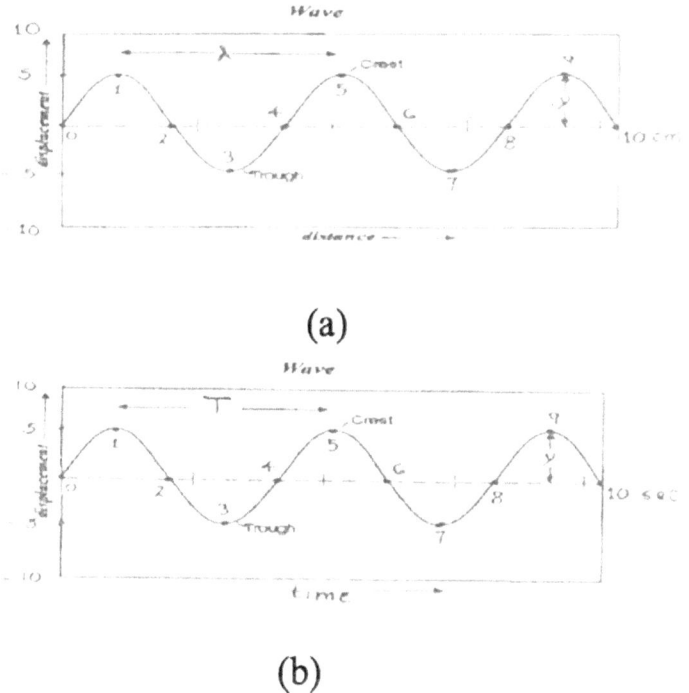

Fig. 2.6 Two representations of a wave's characteristics; both parts have the same numerical values for the axes to simplify the way that a complete cycle or wave can be understood in terms of the wave's length, and how long it takes to complete a cycle

(1) The *Amplitude* of the wave is the maximum displacement from equilibrium, in both cases ±5.
(2) The *wavelength* (λ) from the first diagram is the distance between corresponding points on the wave, in this case 4 cm.
(3) The *Period* (*T*) from the second diagram is the time for the wave or cycle to repeat itself, in this case 4 s.
(4) The *frequency* (*f*) from the second diagram is the number of cycles or waves per unit time passing a given point. Since the basic unit of time is the second, we see that in this case there are two and a half waves or cycles represented over a period of 10 s, so $\dfrac{2.5}{10s}$ = 0.25 Hz. The Hertz (Hz) is the standard unit of frequency and it denotes one cycle or wave per second.

    At this point we can see that the period and the frequency are inversely related, which is generally expressed as $T = \dfrac{1}{f}$.

(5) The *speed* or *velocity*, both terms will be used interchangeably in this book and denoted by the same symbol *v*. Since the general definition of speed is the change in distance over time we can write $v = \dfrac{D}{t}$ and since the change in distance is the wavelength, and the time is the period, we can write $v = \dfrac{\lambda}{T}$ Using $T = \dfrac{1}{f}$ $v = \lambda f$.

    Either highlighted equation can be used to find the wave's speed.

    It can be seen from the figures above that a wave with larger wavelength will also take longer to complete a cycle, thus having a greater period, and a smaller frequency. On the other hand, very short waves have correspondingly small periods and large frequencies.

(6) The *Energy* of a wave will depend on different wave properties. For mechanical waves (regardless of whether they are transverse or longitudinal) the energy depends on their *Amplitude*, and for electromagnetic waves it depends on their *frequency*.

The need for clarity and specificity when using terminology to describe properties of waves cannot be over emphasized. Consider the use of the term "quickness" when referring to the motion of objects; if you are talking about waves you may say that a wave travels quickly and also say that waves are arriving quickly at a point. You are using the same term but you are referring to different properties of waves. The first one is indeed a measure of a wave's speed, but the second is a measure of a wave's frequency.

Exploratory Task

Simple Harmonic Motion using a Pendulum

A pendulum is a convenient device to demonstrate the principles of simple harmonic motion, and the various wave properties introduced above. Its back-and-forth movement can be represented as a wave, where the position of the mass at the end of the pendulum changes from a reference point.

(A) Virtual Part Using the *PhET* **pendulum** simulation software (at http://phet.colorado.edu/index.php), choose a given length and mass for the pendulum. Run the simulation for a small angle (the Amplitude of the oscillations) about 10°. Let the pendulum complete ten (10) complete oscillations (an example of an oscillation is given below).

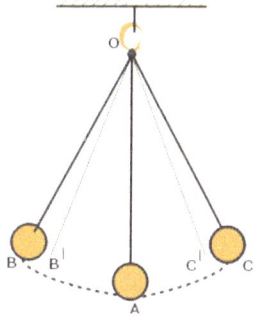

A complete trip or oscillation of the pendulum is the time taken such that if released from B, the pendulum must return to B.

Record the time for the ten oscillations by using the stopwatch, and then divide that time by 10 to obtain the period (T) of the pendulum's motion.

(B) Laboratory Part   Since an oscillating pendulum describes simple harmonic motion, we can use a motion detector or a mobile device with an App to describe the pendulum's motion, and thus determine the period from the graph.

(continued)

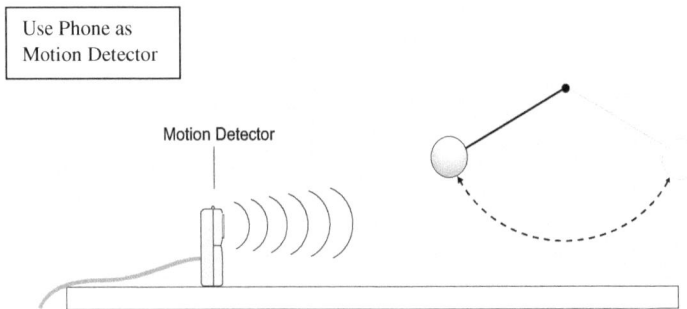

Choose the same mass and length as you did in the simulation, using an
Amplitude of about 10° release the pendulum and collect data with the
phone in the position of the motion detector. Capture the motion as a graph
and determine the period from the graph by dividing the total time of
motion by the number of peaks.

Determine the % error between the period from the simulation and your
experimentally determined one above.

$$\%error = \left| \frac{T\,simulation - T\,experiment}{T\,simulation} \right| \times 100$$

$$\%error = \cdots$$

Discuss your results by pointing out the likely reasons for your calculated
errors.

Exercises

1. (A) What letters can be used to represent the Amplitude of the waves
   shown in Fig. 2.6?
   (B) Corresponding points for the wavelength and the period are sets of
   two numbers. Besides (0–4) cm, what other sets can be used?
   (C) What is the speed of the above waves?

2. Describe the wave shown below in terms of the property that is changing

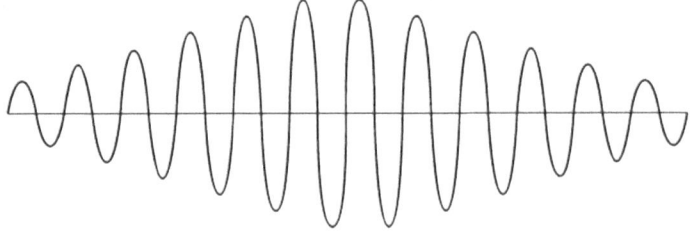

(continued)

3. In all diagrams the dashed line represents the equilibrium position

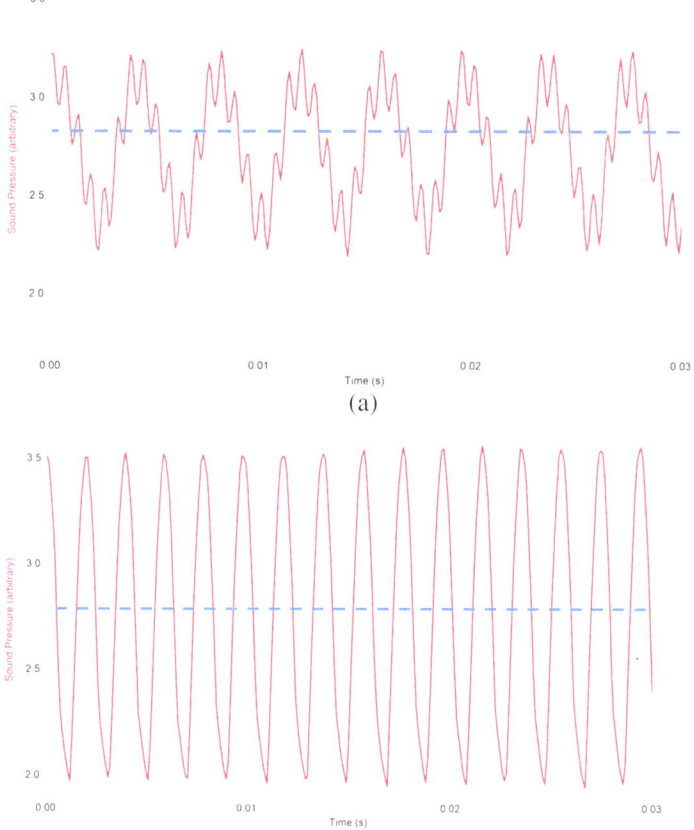

(a)

(b)

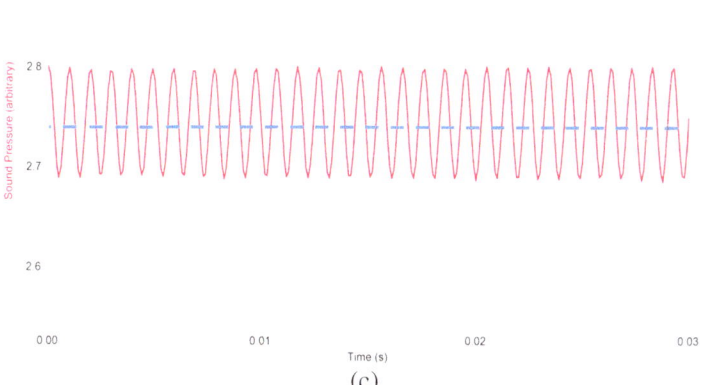

(c)

(continued)

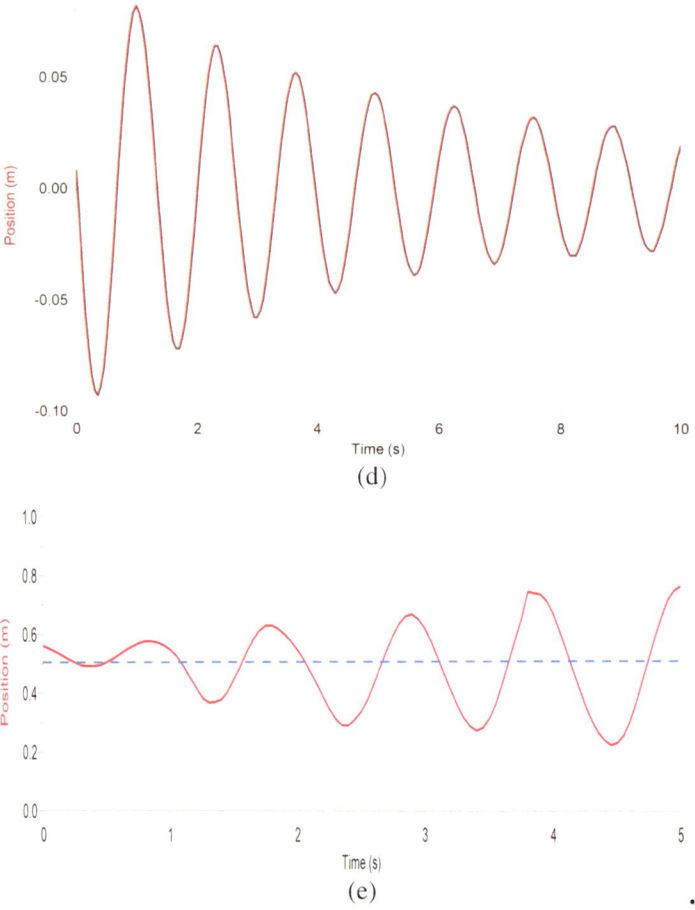

(d)

(e)

Rank the above 5 waves from highest to lowest in terms of:

(A) Frequency
(B) Amplitude
(C) How do (d) and (e) differ from the others?
(D) How does (e) differ from (d)?

   (*) Note—Use Amplitude and Frequency in (C) and (D).

(continued)

Exercise

Compare the three situations shown below; in each case the spring has been moving back and forth after a force set it in motion. Assume that K_1 and K_2 are equal, K_3 is larger than both of them, and X_1 and X_3 are equal, but X_2 is shorter than them.

Match the graph shown on the right column with the appropriate diagram shown on the left one.

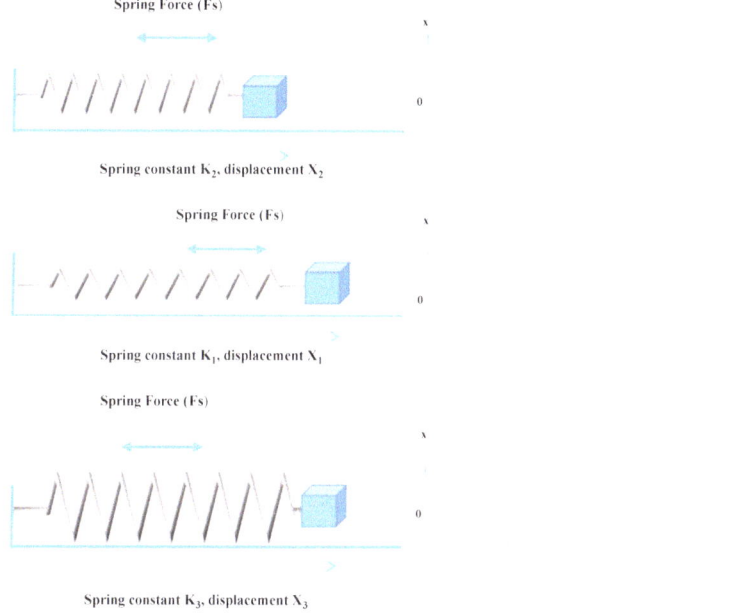

Spring Force (Fs)

Spring constant K_2, displacement X_2

Spring Force (Fs)

Spring constant K_1, displacement X_1

Spring Force (Fs)

Spring constant K_3, displacement X_3

HINT: The displacement of the spring is equivalent to the Amplitude of the wave, and the greater the value of K the stiffer the spring is and so the lower will be the frequency of the wave.

(continued)

Exploratory Task

Does the speed of a wave in a rope or a string depend on how fast the oscillations are created?

ANSWER:

What are your reasons for the answer?

Now test your answer by using a simulation at http://phet.colorado.edu/index.php.

Choose "Wave on a string" from the available choices. With the simulation open, make sure it looks like the diagram below.

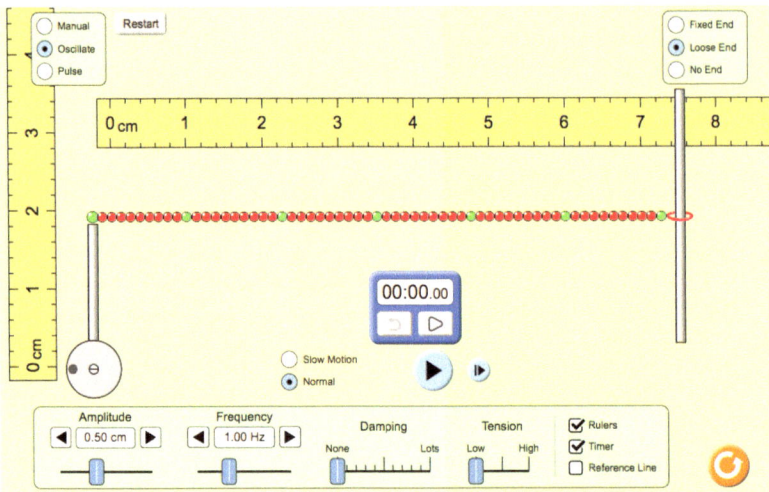

The horizontal ruler can be moved to determine the wavelength, while the vertical one stays put.

- Click *as simultaneously as possible* the Oscillate and Play/Start buttons and determine how long it takes for the wave to reach the ring at the end. Stop the timer and record the value. Determine the wavelength by placing the horizontal ruler so that it measures the distance between peaks. Repeat for "Frequency" (the equivalent of how fast the oscillations are set up) values of 2.0 and 3.0.
- Now select "Frequency" back to 1.0 and then change the tension to the middle of the scale; repeat the measurement of the time. Finally, change the tension to high and repeat once more, then fill in the table.

(continued)

(*) The speed can be calculated by multiplying the wavelength by the frequency ($v = \lambda\, f$)

Frequency (Hertz)	Tension	Timer Reading (seconds)	Wavelength (centimeters)	(*) Speed (cm/sec)
1.0	low			
2.0	low			
3.0	low			
1.0	middle			
1.0	high			

1 What does the table suggest for
   (a) The time taken for the waves to travel?
   (b) The speed of the waves?

2 (a) What did you notice about the wavelengths as you made the changes?
   (b) Can you offer reasons for the answer in a)?

Other properties of the material components determine the speed of a wave. In the case of a rope or string their tension and thickness have effects such that a greater tension results in higher wave speed, and greater mass per unit length (thickness) yields a lower speed. In the case of sound, the temperature of the medium through which the sound propagates and its density have similar effects on the speed of sound. However, once these properties are fixed or maintained so that they are constant, as waves are generated and propagate, the above list of six (6) properties is sufficient to account for the properties of wave motion.

The speed of a wave determines how quickly the particles or components of the wave execute the motion caused by the source of the wave. However, the speed resulting from the motion of the components isn't the same as the wave's speed. In the stadium example from the previous chapter, how quickly you rise or lean against your neighbor will not affect how quickly the created pulse or wave moves through the stands. If anything, the faster your motion in either case the greater the frequency of the wave since it would take you a shorter amount of time to complete your motion. Of course, the wavelength will also change accordingly, as shown by the preceding simulation. We ought to clarify that in this case, a person's motion can be considered as both the source of the waves and the material through which they travel. As a wave of this type travels around the stadium if some people begin to move differently, this will of course lead to the generation of a new wave.

Application to Light

The speed of light as well as all electromagnetic waves has an interesting and unique property that can be seen from the entire spectrum in Fig. 2.7, it is constant.

Activity
If you draw a vertical line that connects the horizontal wavelength and frequency lines (a dashed one is drawn going through the FM radio tower) regardless of where in the spectrum you do this, you will find the same thing. Namely, if you approximate the value between the numbers along the wavelength (they all vary from 1 to 10) and multiply it by the value of the frequency (remember $v = \lambda\, f$), you will get $300{,}000{,}000 \ \frac{meters}{second}$ or 3.0×10^8 $\frac{meters}{second}$ as we shall see later in the chapter, when multiplying numbers in scientific notation one adds the powers or exponents. In the case of the dashed line, the wavelength is approximately 3 m, and the frequency is approximately $10^8 \ \frac{waves}{second}$, so their multiplication yields the value of $3.0 \times 10^8 \ \frac{meters}{second}$.

This highlights the fact that all electromagnetic waves have the same speed.

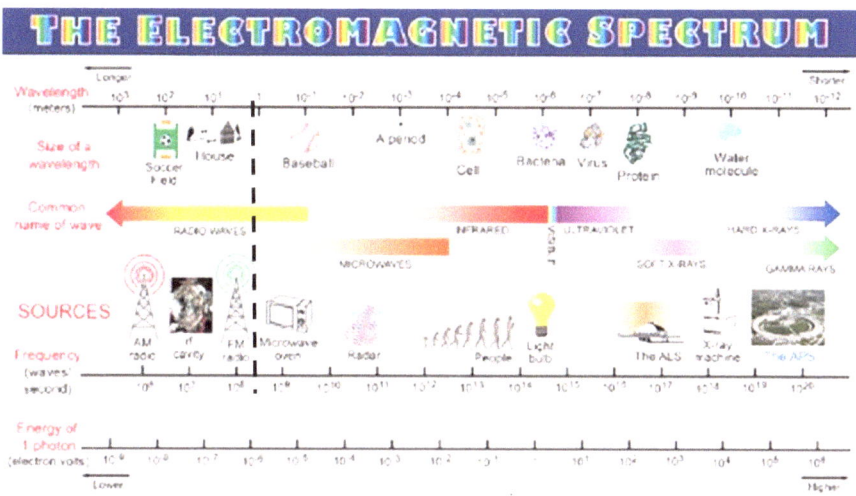

Fig. 2.7 The electromagnetic spectrum (credit *Argonne National Laboratory*)—The diagram shows the visible part as well as those other regions that are invisible to us

Another interesting characteristic of regarding electromagnetic waves as radiation is that they take many forms or types, from heat to highly penetrating X-rays and gamma rays, the shortest waves in the spectrum. Additionally, all objects radiate energy depending on their temperature, and this energy can be distributed across different parts of the spectrum. Figure 2.8 illustrates how various objects can emit different forms of radiation.

Figure 2.8 shows the curve displayed by all the radiation emitted by an object, also known as the blackbody radiation curve. The temperature of the thermometer is shown in absolute or degrees Kelvin ($100\ ^\circ C \approx 373$ K). In part (a) the radiation is that emitted by the sun at 5778 K, where the curve peaks in the yellow part of the visible spectrum, where the sun radiates the most energy. The three letters (B, G, and R) that stand for the primary colors Blue, Green, and Red are shown next to the white symbol indicating that the entire spectrum is visible. The curve also hovers over regions with wavelengths shorter than 0.4 μm (the ultraviolet) and longer than 0.7 μm (the infrared), indicating that the sun also radiates in those invisible parts of the spectrum beyond the visible range.

In part (b) the radiation emitted by a campfire at 1500 K shows the curve peaking at around 2 μm way beyond the visible range, although there is a tiny amount under the curve towards the red end of the visible range, as indicated by the Red letter being the only one highlighted above the curve. This means that a campfire releases energy that is overwhelmingly along the infrared and longer regions of the spectrum, with very little being visible.

Finally in part (c) the radiation emitted by mammals at room temperature around 300 K is shown peaking at 10 μm and being completely outside the visible region of the spectrum. The curve only begins to deviate from zero intensity at about 4 μm, and the line representing the visible region is shown all the way on the left of the

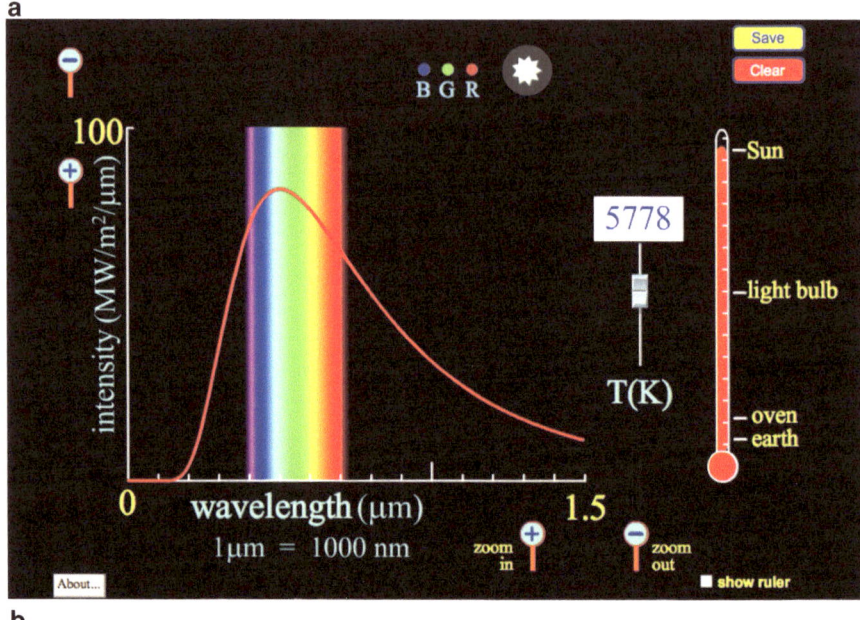

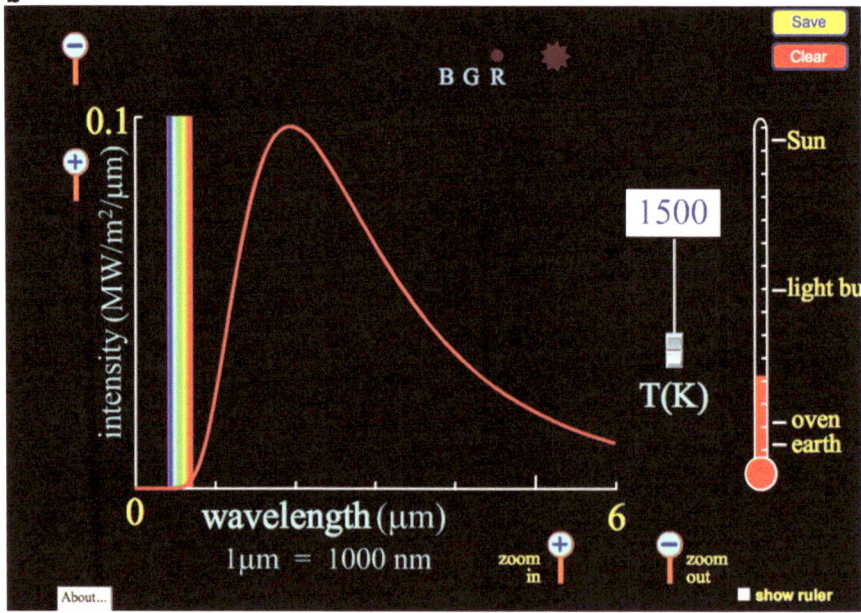

Fig. 2.8 (**a**) Radiation emitted by the Sun, with an approximate temperature of 5778 K in absolute units. The radiation or amount of energy emitted peaks in the yellow of the visible part of the spectrum, but there is also radiation in the ultraviolet part (shorter wavelengths), as well as in the infrared part (the longer wavelengths). (**b**) Radiation emitted by a campfire at around 1500 K, the radiation peaks beyond the visible part of the spectrum (in the infrared). There is very little radiation in the visible part, with most radiation in the infrared or heat part of the spectrum.

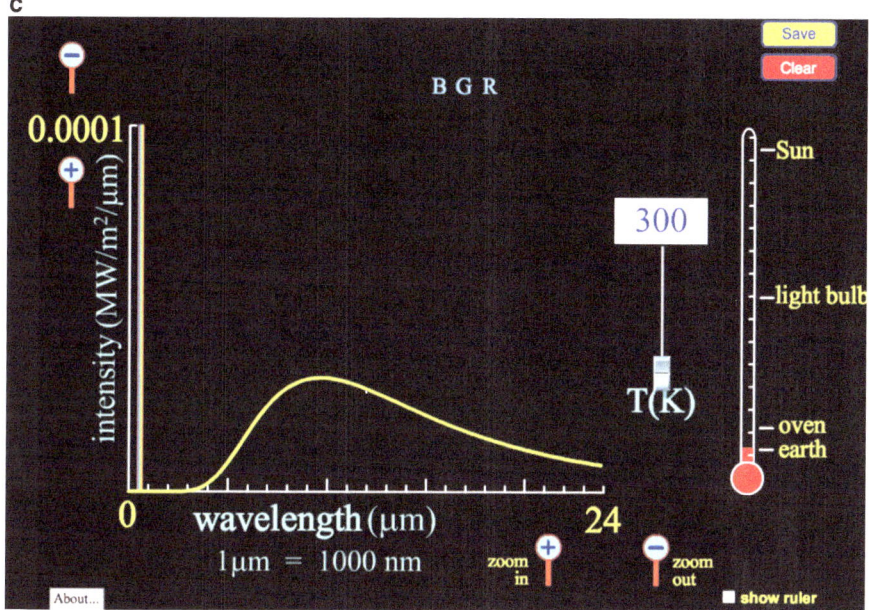

Fig. 2.8 (continued) (**c**) Radiation emitted by mammals at around 300 K, which is roughly room temperature. This illustrates that particularly in the dark, most of the radiation we emit is in the infrared part, which makes us invisible unless light reflects off our bodies

graph. The figure explains why we are invisible in total darkness (none of the letters above the curve are highlighted), although we radiate quite a lot of energy in the infrared region.

Exercise
Use Fig. 2.8a to explain why you shouldn't look at the sun without eye protection during a total eclipse when the disk will be blocked and no visible radiation gets to your eyes.

Application to Sound

Figure 2.9 shows the most commonly used representation of a wave; although as introduced in the second type of wave classification, longitudinal waves are the result of motions different from the ones that the figure represents.

Figure 2.10 represents the vibrations produced by a tuning fork where the surrounding air turns into alternating bands of compression and expansion. The dark bands correspond to light dashed arrows denoting maximum Amplitude of the wave representation, and the bright bands correspond to black dashed arrows indicating

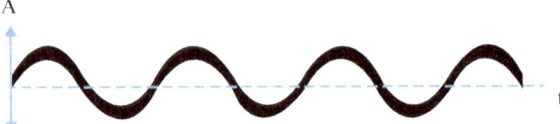

Fig. 2.9 A sine wave (where the oscillations begin with zero Amplitude at the beginning) is commonly used to represent oscillatory motion. The highest and lowest points represent the Amplitude, and the middle (*dashed line*) represents the points where the oscillations return to the equilibrium/initial position

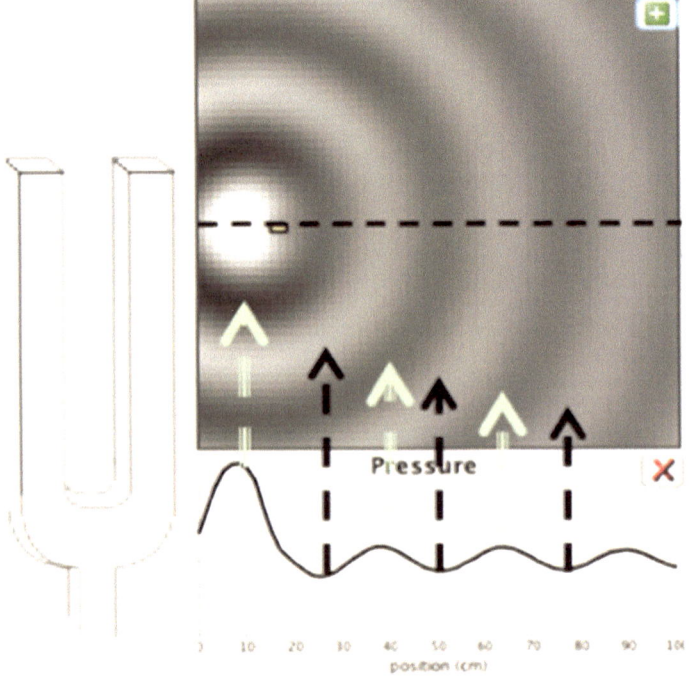

Fig. 2.10 The figure is a representation of a longitudinal wave resulting from the motion of a tuning fork. Once struck, its vibrations set the surrounding air in motion and there are regions where the air is compressed (*the dark bands*), as well as regions where the air expands (*the bright bands in between*). The sine curve underneath represents the alternative progression of the movements, indicating that as the vibrations spread out they lose energy

minimum values on the wave. Notice how the loss of energy is shown by the decreasing intensity of the bands corresponding to decreasing Amplitude of the wave.

The representation of longitudinal waves, particularly sound, requires more detailed explanation since these differ from transverse waves. With transverse waves, the motion of the components and their representation in terms of the list of five properties above are similar. However, in the case of sound the oscillations of the air and their properties are different from the way waves were represented above. Figure 2.11 is an attempt to clarify these differences.

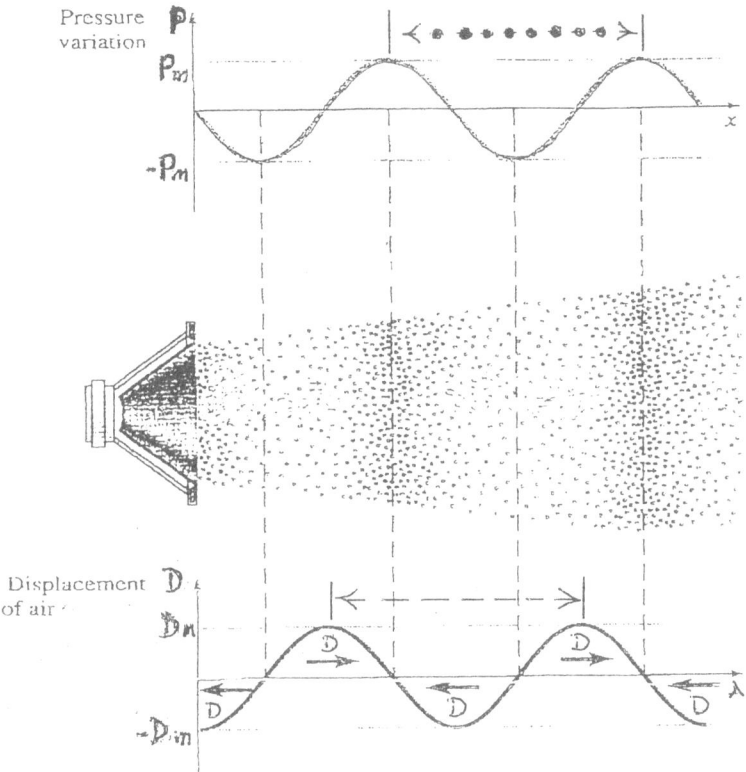

Fig. 2.11 An expanded view of the air motions when sound waves propagate through it. The darker regions represent compression and the lighter ones represent expansion or rarefaction. Two graphs are used to indicate the variations in pressure and displacement or movement of the air. ± Pm and ±Sm represent the points of maximum value for both pressure and air movement. Note that when the air movement is greatest (*bottom graph*), the air pressure is zero (*top graph*). Additionally, as the air movement is in both directions they give rise to the compression and expansion; by contrast the pressure variation is from zero to a maximum value, even if shown as both positive and negative maxima

In general, the compression and rarefaction of longitudinal waves occur at the locations where the medium displacement is zero, as can be seen from the graphs above. Therefore, we should be able to correlate the actual air motions and lack thereof, as well as the variations in air pressure to the wavelike representation, which looks like that of a transverse wave as well. The main difference is that while the wavelength can be obtained from the figure by using either the dotted double arrow for pressure, or the slashed double arrow for air displacement, the Amplitude would have to be indicated differently than what the two graphs show. A larger Amplitude would consist of more dots (darker) in the regions of compression, and fewer ones (lighter) in the regions of expansion or rarefaction, as shown in Fig. 2.12.

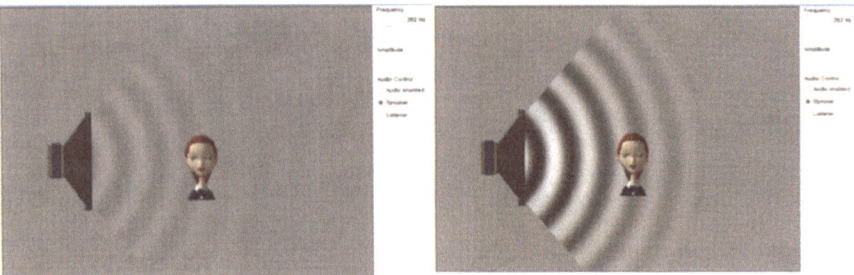

Fig. 2.12 A sound wave produced by the loudspeaker on the left of each diagram can be shown to have a given frequency and wavelength, and yet look different since the Amplitude changes. Note that the left diagram shows a fainter set of wave fronts as compared to the right one where they are darker and brighter, respectively, due to the wave having a larger Amplitude

There are many other situations where phenomena can be described in terms of wave-like properties. One of the more common ones is the motion of a simple pendulum where its back-and-forth movement has such characteristics that they can be described with the properties listed above. The motions of buildings and bridges, of the ground in seismic, and of the air in weather-related events can also be understood in terms of wave characteristics.

A structure like a bridge or a building can move in complicated ways, but its overall motion can often be described as that of an object obeying the rules of simple waves, with given properties like those described in this chapter. For example, the period (T) for a 10-story building is approximately 1/2 s. For the World Trade Center [1350 ft high] it was approximately 10 s, and the Amplitude of its oscillations (if you can believe that a structure like that would sway back and forth!) was about 3 ft at the top. Information like that can be very useful for analyzing and understanding their behavior, particularly when they vibrate under the effect of winds, and earthquake waves.

Exercise

There are instances of people feeling sick in workplaces where machinery generate waves of frequencies that can interact with the body's internal functions. In the case of a building that oscillates back and forth when acted by external vibrations and waves such as high winds:

(A) Where would people feel the greatest effect of its swaying back and forth like a pendulum, at the lower or higher floors?
(B) What property of a wave would that motion represent?
(C) How comfortable would you feel living in a skyscraper knowing that in order to prevent the building from swaying too much under the effects of strong winds, an 800 ton weight is used in a special room near the top of the building?
(D) In order to further explore the above situation read the article "Height Meets Heft" (The New York Times, Real Estate section. Sunday August 9, 2015). What concepts from your understanding of waves do you find discussed in the article?

The last paragraph illustrates examples of a phenomenon particular to waves known as *resonance*. Objects that experience vibrations can oscillate with a variety of frequencies; however, they tend to have a dominant or resonant frequency that depends on a number of factors, such as their length, thickness, and other properties like their structural composition. Whenever an object is set in motion by a force that repeats its effect on the object periodically, if the frequency with which the force changes matches the resonant frequency of the object, the Amplitude, and consequently the energy of the oscillations will be at a maximum. Structures like buildings and bridges are protected by isolation points or damping features, from vibrations caused by external forces that can result in the structures themselves vibrating at their resonant frequencies.

Exploratory Task

The concept of resonance can be demonstrated by the use of the same simulation that we used in previous tasks "Wave on a String" available at http://phet. colorado.edu/index.php.

When a wave initially sent from the left (the rotating oscillator) reflects from the Fixed End, we now have two waves interacting. With the simulation open, make sure it looks like the diagram below.

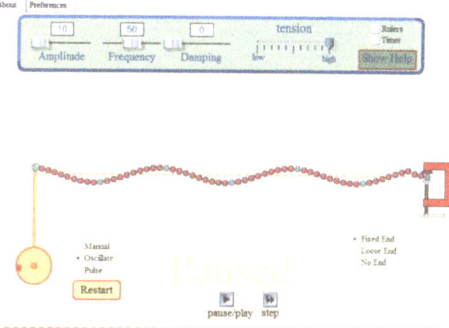

- Click on Oscillate (after you select the required values for Amplitude, Frequency, and Damping, as well as high tension, and Fixed End) and observe the wave for 30 s. Stop the simulation and describe the situation.
- Choose "Loose End" and repeat, what do you observe?
- Finally change the Amplitude to 50, the Damping to 10, and go back to "Fixed End". Run the simulation and describe what you observe.

The various characteristics and properties of waves introduced in this chapter can be used to practice and familiarize ourselves more with some of the real-world applications of concepts in light and sound. In terms of human perception there are two sets of ranges for which phenomena are both audible and visible.

For sound the range of audible frequencies is roughly (20–20,000) Hz, and for light the visible range is usually given in terms of wavelengths, roughly (400–750) nm or $(4.0–7.5) \times 10^{-7}$ m. There are of course other waves beyond those ranges, but they are not perceived by humans. These regions are respectively known as the *ultrasonic* (greater than 20,000 Hz), *infrasonic* (less than 20 Hz) for sound; correspondingly there is the *ultraviolet* (less than 4.0×10^{-7} m) and *infrared* (greater than 7.5×10^{-7} m) for light. The terminology doesn't change for sound, but it does for light due to the different colors perceived at both extremes of the visible spectrum.

In general, *ultra* means "above the human perception range" and *infra* means "below the human perception range." Don't be confused by the distinction between the ranges for sound and light; the term *ultra* is used not only for very large sound frequencies, but also for very short light wavelengths (which in turn correspond to very large light frequencies). The same thing applies to the term *infra* when referring to very low values. This is an unfortunate result of the way the two ranges are usually described, with one in terms of frequencies and the other in terms of wavelengths. Nevertheless, there should be no confusion if you understand the inverse relationship between frequency and wavelength.

Experimental Task
Determine your heart rate or pulse (in beats per minute) by finding a place where it is easiest to measure, such as on your neck or at the wrist. Once found, count the number of beats in 1 min. What are:

(A) The period T (Hint: divide the number of beats by 60 s)
(B) The frequency f
(C) What do you hear when your head lies on a pillow and you feel pressure changes against your ear that repeat themselves?

Activity

The following graphs represent the waves produced by the sound of a piano and an organ.

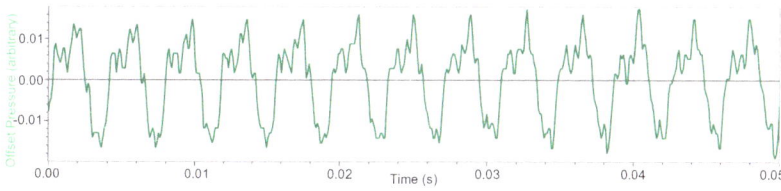

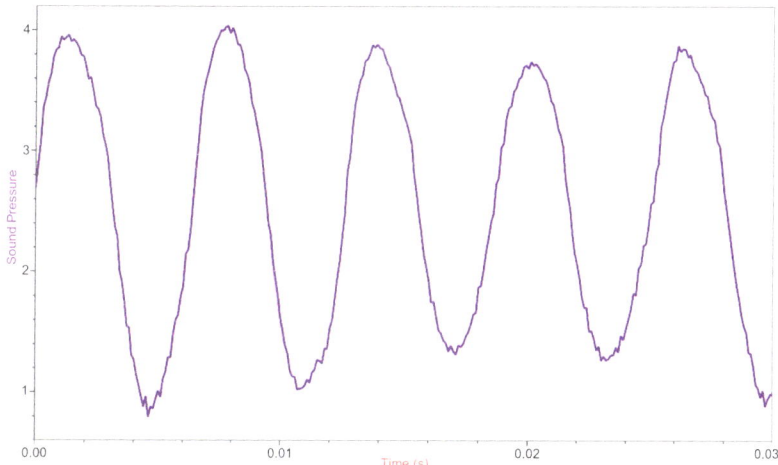

(1) Can you predict from the graphs which instrument has a higher frequency (pitch)? How?
(2) Determine the respective periods from each graph by dividing the total time by the number of waves (remember to use corresponding points to find the completed cycles or waves).
(3) Determine the respective frequencies from the periods.

Exploratory Task
Investigation of the frequency of vibrations and vibrating objects

(Using a microphone and an interface to display the frequencies of air vibrations and those of a glass container).

(continued)

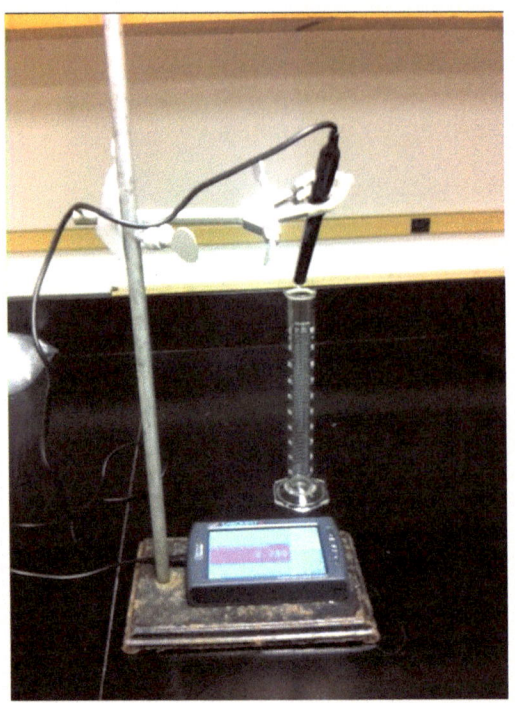

Prediction: Draw a graph of what you think the relationship is between the pitch (frequency) of vibrations and (a) changes in the length of the column of air in a glass container as you blow across its top, (b) changes in the amount of water poured into the container as you strike it with the eraser end of a pencil or some other soft object.

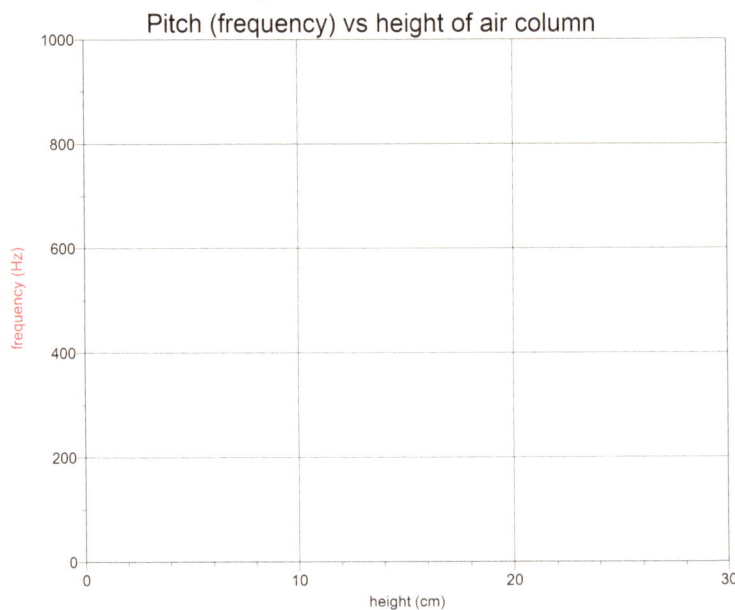

a) **Blowing across the top of a glass container**

(continued)

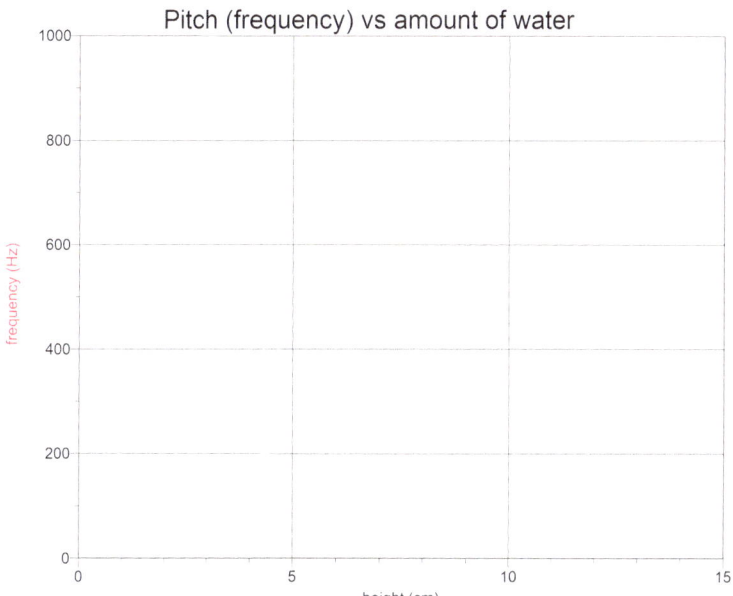

b) **striking a glass container with a pencil**

NOTE: In both cases you may want to make the first observation when the glass container is empty.

1. Air Vibrations

In the first part we investigate the vibrations produced by blowing across an air column and determining the relationship between the length of the column and the frequency of the vibrations. Use a glass container and pour a certain amount of water into it, blow across the top, and notice the pitch for each amount of water added; for each amount of water added, what relationship do you determine between the height of the air column and the pitch of the sound produced when you blow across the top of the container?

Now repeat the activity, but use a microphone connected to the Lab Quest interface to measure the frequency for several water levels. Repeat until you get a "clean" display (the dominant frequency) on the Lab Quest screen. Fill in the table and plot the relationship between the frequency and the air column (*remember to subtract the height of water from the top of the container to determine the height*).

To get the frequency you must first determine the period (T) of the signals displayed on the Lab Quest/computer screen, then use $f = 1/T$

NOTE: there should be around 10–11 peaks for a .03 s time range with an empty 100 ml graduated cylinder.

Water level (cm)	Air column height (cm)	Frequency (Hz)

What do you notice from the table that resembles what you observed when you listened to the pitch as you added water to the glass bottle and blew across its top?

2. Glass Vibrations

In the second part we investigate the vibrations of a glass container by varying the amount of water poured into it. For each trial pour an amount of water into the container, strike it gently but sharply with the eraser end of the pencil and notice the pitch, what do you observe between the pitch and the changes in the amount of water?

Again, for each water level, strike the container with the rubber part—the eraser end of the pencil or soft object and determine the frequency of the vibrations with the microphone. Fill in the table and plot the data. (*In this case we don't subtract the water level from the top of the container since the vibrations are produced by the glass as it is filled with water.*)

Water level (cm)	Frequency (Hz)

What do you notice from the table that resembles what you observed when you listened to the pitch as you added water to the container?

Attach your graphs and a section where you compare the results of both activities, along with a discussion of similarities and differences between them. How do the graphs compare to your predicted ones?

Scientific Notation

We use scientific notation to express very small and very large numbers; the idea is to keep track of the location of the decimal place.

To write 100 in scientific notation, we realize that the decimal place is after the last zero, so we express the number or the base with an exponent or power that accounts for the number of places one has moved either to the left (positive exponent/power) or to the right (negative exponent/power). Hence 100 can be written as 1.0×10^2 since we have moved the decimal place twice to the left. Correspondingly, $\frac{1}{100}$ or 0.01 can be written as 1.0×10^{-2} since we have moved the decimal place twice to the right.

Adding and subtracting numbers in scientific notation can be made easier when they all have the same exponent or power; multiplying numbers in scientific notation requires that the numbers/bases themselves be multiplied but the exponents/powers be added; division requires the numbers to be divided but the exponent/power in the denominator is subtracted from that in the numerator.

Exercises
1. Add 1.3×10^3 to 3.7×10^4 (Hint: one of the numbers should be changed so that the exponents/powers are the same)
2. $(2.0 \times 10^{-2} - 3.0 \times 10^{-2})$
3. **(a)** $(5.2 \times 10^{-3})(6.0 \times 10^2)$ **(b)** $(2.0 \times 10^5)(8.0 \times 10^1)$
4. **(a)** $5.0 \times 10^4/2.5 \times 10^3$ **(b)** $9.0 \times 10^{-2}/4.0 \times 10^{-6}$

Units

We use the decimal/metric system of units where any of the measurements can be expressed in the respective units, as well as their fractions and multiples in terms of prefixes (Table 2.1).

Table 2.1 Most commonly used units and notation in our study of wave properties

Length	Mass	Time	Frequency	Prefix	Factor	Scientific notation
↓	↓	↓	↓	Pico	0.000000000001	1.0×10^{-12}
Meter (m)	Gram (g)	Second (s)	Hertz (Hz)	Nano	0.000000001	1.0×10^{-9}
				Micro	0.000001	1.0×10^{-6}
				Milli	0.001	1.0×10^{-3}
				Centi	0.01	1.0×10^{-2}
				Kilo	1000	1.0×10^3
				Mega	1,000,000	1.0×10^6
				Giga	1,000,000,000	1.0×10^9
				Tera	1,000,000,000,000	1.0×10^{12}

Exploratory Task Using Both Light and Sound

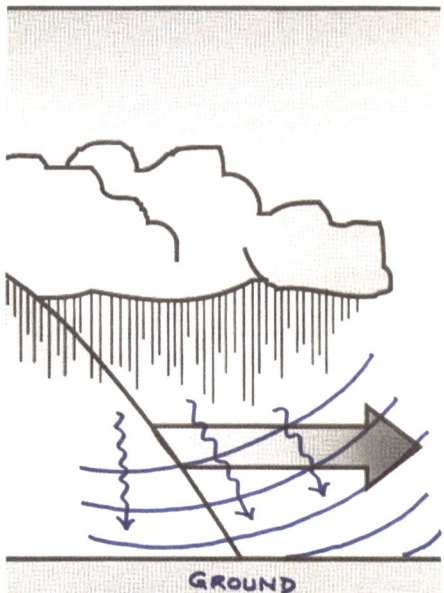

Have you heard of the five second rule when witnessing a thunderstorm? What do you think it means?

It has to do with predicting when thunder will occur once you have seen the lightning during a storm; have you noticed that one always sees the lighting before hearing the thunder?

We can demonstrate it by using the considerable difference between the speed of light and that of sound. The speed of light is about a million times greater than that of sound (300 million meters/second, as opposed to roughly 340 m/s); therefore, for light and sound travelling a given distance through the air, the time difference is in the order of a millionth of a second.

Sound travels through the air with an approximate speed value of 340 m/s, so in 1 s the sound produced by the thunder will travel about 340 m. Now, a mile is roughly 1600 m and $\frac{1600\,\text{m}}{340\,\text{m}} = 4.7 \approx 5$ which means that a mile is approximately 5 times longer than the distance sound travels in 1 s, 340 m.

Therefore, the five (5) second rule states that when you see lightning, count to 5 s. If you hear the thunder before you finish counting, then the lightning took place less than a mile away, and if the thunder is heard after you counted to 5 s the lightning took place farther than a mile away.

Task: Try this next time you witness a thunderstorm (of course making your observations indoors), and determine how accurate the five second rule

(continued)

is; make sure you repeat the observation several times to find the shortest and
longest distances where the lightning occurs during your observations.

Exercise

Suppose you see lightning strike the ground a mile away:

(a) How long does it take the light to reach you?

Hint: Use the following

$$\text{Speed } (v) = \frac{\text{Distance } (D)}{\text{Time } (t)} \quad \text{therefore solve for the time } t = \frac{D}{v} \text{ and use}$$

$$v = c = 3.0 \times 10^8 \frac{\text{meters}}{\text{second}}$$

c is the symbol used for the speed of light through the air (which is an
approximation anyway, but works reasonably well). The distance D is
1609 m (the equivalent to a mile).

(b) How much quicker than sound is the time for light to reach you?

Chapter 3
Reflection

The interaction of waves with objects generally results in both reflection and absorption depending on the types of objects the waves encounter. Similarly to what we saw in Chap. 2 when a pulse travelling through a rope or string reached its end and was in turn reflected upside down or upright, depending on whether the end was flexible or fixed.

The objects chosen to explore the reflection of waves are extended so that they appear as surfaces, and irregular or rough, as well as regular or smooth surfaces are considered; however, the more extensive applications are found in the case of smooth surfaces since in this case the waves remain fairly focused. In the case of reflection from rough surfaces, the waves tend to disperse and scatter and so their study gets more complicated.

Reflection of Light Waves

There are two types of reflection: regular or *specular* reflection from smooth surfaces, and *diffuse* or irregular reflection from rough surfaces. In order to consider both types we need to use a model of wave propagation that simplifies the way they interact with surfaces. This model was first developed by the Dutch physicist Christiaan Huygens in the seventeenth century.

Huygens' principle can be demonstrated by the use of the concentric circles in Fig. 3.1b. The circles represent spreading wave fronts away from the source S in all directions. Huygens' principle consists of the assumption that each point on a wave front in turn generates secondary wave fronts.

Figure 3.2a shows the wave fronts spreading from a single source; (b) shows those emanating from two sources; (c) shows the pattern produced by three sources, and (d) that produced by five sources. Note that as the number of points increases the secondary wave fronts they generate become superimposed as they spread out, resulting in a pattern that will eventually become a continuous wave front when the

© Springer International Publishing Switzerland 2017
F. Espinoza, *Wave Motion as Inquiry*, DOI 10.1007/978-3-319-45758-1_3

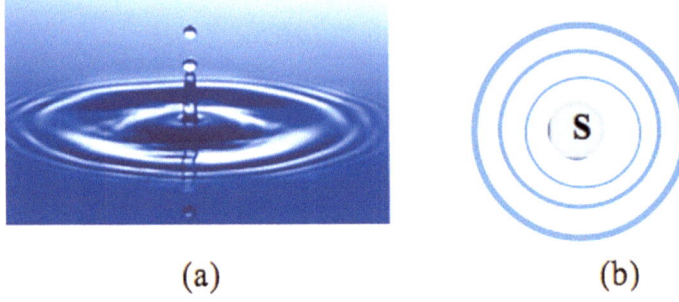

Fig. 3.1 A representation of a source of waves such as the object that causes the ripples in (**a**) by falling on the still water surface. In part (**b**) the perspective is that of a source S that radiates light in all directions as shown by the concentric circles surrounding it. In both cases the circles represent the ripples spreading out from the source

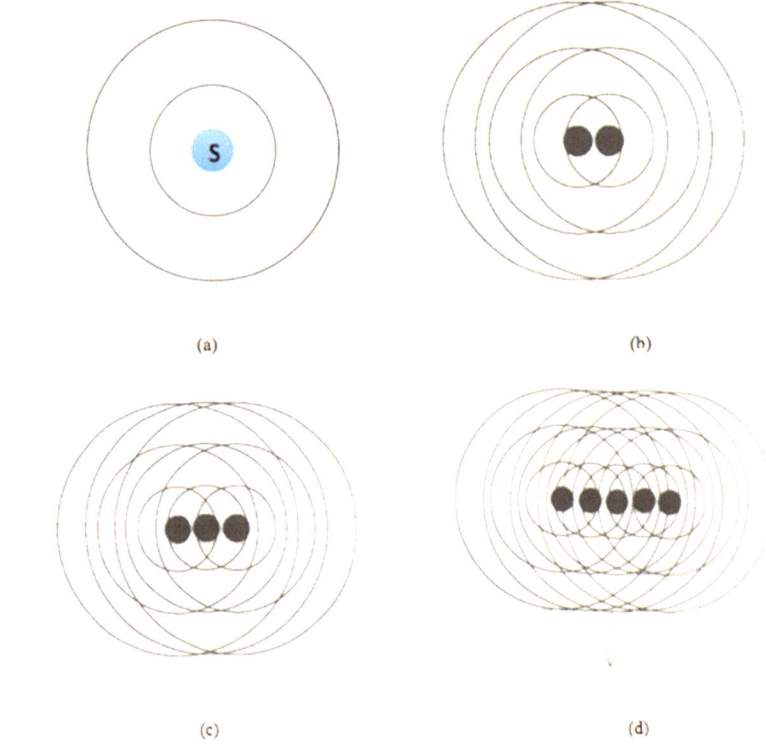

Fig. 3.2 Evolution of the contribution of secondary wave fronts to the formation of a pattern that propagates radially from a spherical surface. After a certain time the new pattern describes a position of the wave front resulting from an infinitely large number of points, as a surface tangent to the contributing secondary waves. That pattern (a new wave front) can then be replaced by a line that is always perpendicular to it

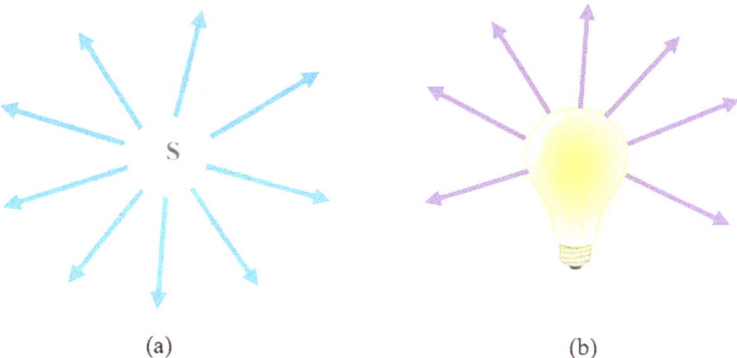

(a)   (b)

Fig. 3.3 Gradual representation of the spreading of light from a source S where the concentric circles (the wave fronts) are supplanted by *arrows* representing rays of light. Note that in (**a**) the rays are always perpendicular to the wave fronts, and that only a few rays are drawn as there are in principle an infinite number of these. Using the convention we can visualize how light spreads from a source, such as a light bulb in (**b**)

number of points becomes infinitely large. This pattern, the dotted line in (d) is tangent to the contributing secondary waves. Hence the light that emanates from a spherical surface can be represented by the arrow in (d) that will be everywhere perpendicular to the wave front (the tangent dashed line) created by the infinite number of points lying on the surface; in other words, by a *ray* of light. This is illustrated in Fig. 3.3, where the arrows represent rays of light emanating from the surface S.

The reflection of light from surfaces can now be strictly considered by regarding the light as propagating through the various materials in the form of rays. When a source such as the sun is used, the rays emanating from it are regarded as being parallel to each other (forming a beam), since the sun is considered to be infinitely far away compared to the other distances involved.

This assumption can also be used for ordinary light sources, such as light bulbs, whenever they are sufficiently far away from the objects and surfaces the light interacts with. The fact that light appears to propagate ordinarily in straight lines (think search beams) confirms such an assumption.

With the use of light rays, as shown in Fig. 3.4 we can tackle the many situations where light is reflected from surfaces. As we said before, and as is shown in Fig. 3.5 reflection from surfaces that are very irregular or rough results in the reflected light being dispersed, while reflection from smooth or regular surfaces leads to a condition where a law can be stated. This is the *law of reflection*: **when a light ray strikes a smooth surface, the angle of incidence is equal to the angle of reflection** (both measured from an imaginary line, called the normal).

Using the law of reflection we can explore the properties of mirrors, we begin by looking at a plane mirror.

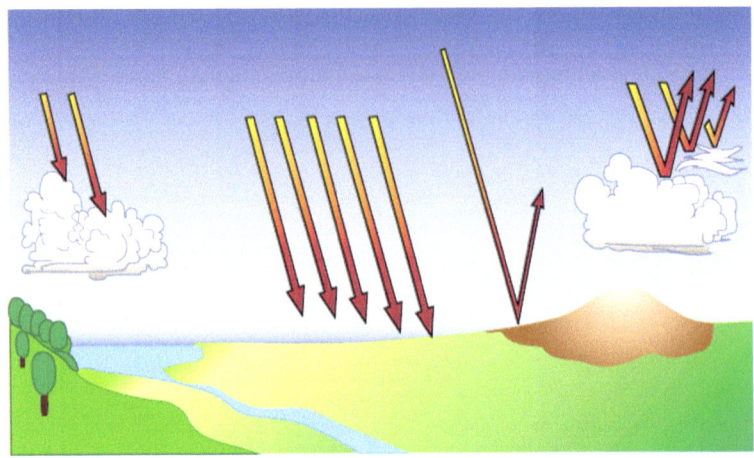

Fig. 3.4 The use of rays greatly simplifies the description of the way light interacts with objects and surfaces. In the figure, direct as well as reflected sunlight can be represented by light rays

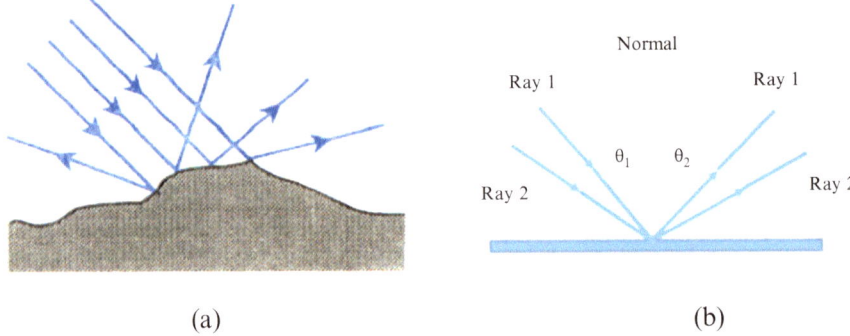

(a)   (b)

Fig. 3.5 (**a**) Diffuse or irregular reflection, (**b**) specular or regular reflection. In both cases the *arrows* pointing to the surface are called incident rays, and those pointing away from it are called reflected rays. The law of reflection applies only to (**b**) where the angles θ_1 (called the angle of incidence) and θ_2 (called the angle of reflection) are equal. Two rays are shown in (**b**) and the condition is applied to Ray 1, but it also applies to Ray 2, in which case the angles measured would be different from those for Ray 1, but still would be equal to each other

Activity

The oldest type of mirror found in museum collections of ancient artifacts from the earliest civilizations was made of metal. If you shine a light source such as a flashlight on a metal cabinet and then shine it on a wall, you can see the difference in the reflected light. Why do you see an image on the metal but not on the wall if the light is reflected from both surfaces?

Plane Mirrors

Mirrors have been made for centuries by coating a glass surface with aluminum or silver. A plane mirror is the simplest type and we establish a few conditions so that we can investigate the properties of images produced by it.

Figure 3.6 illustrates the properties assigned to objects and images. All light rays originate at the object and they become the incident rays upon striking the mirror. The thin dashed lines perpendicular to the mirror represent the normal lines drawn at the points where the incident rays strike the mirror. The continued dashed lines are the projections of the reflected rays behind the mirror. It is important to recognize that only solid lines are considered to be real, dashed ones are not. As can be seen from the figure if one measures the distance from the mirror to the tip of the black arrow and to point P, plane mirrors have the unique properties that the object and image size or height, as well as their distances from the mirror, are always equal.

In addition to the above properties, we always assume that the intersections formed by solid rays whether they emanate from an object or converge at an image result in real objects and images. In contrast to situations like the one at point P, the intersection of projected rays results in virtual objects and images. The deciding physical characteristic of real objects and images is that they can be projected onto a screen or other smooth surface, whereas virtual ones cannot; although they still exist in the sense of being perceived.

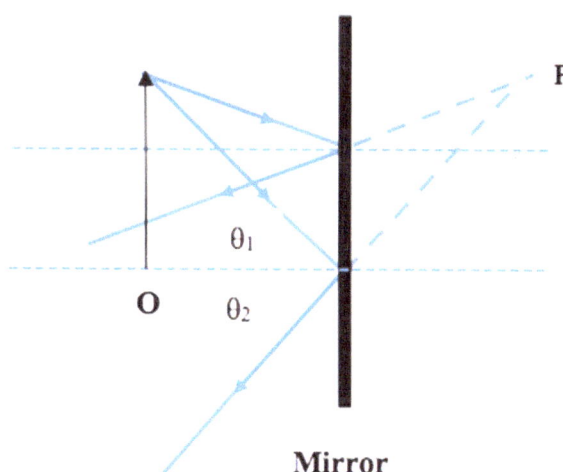

Fig. 3.6 A plane mirror showing two rays emanating from the tip of the *black arrow* drawn from point O that represents the object, with its height or size determined by the length of the arrow. Both rays obey the law of reflection, although only one case is shown where the two angles θ_1 and θ_2 are equal. The two reflected rays spread out, but their projections (*the continued dashed lines* behind the mirror) appear to cross at point P where the image forms

Conceptual Task

Using Fig. 3.6 to represent the cabinet mirror in your bathroom, with the arrow at 0 representing where you are in front of the mirror, when you look at your reflection what do you notice about your distance from the mirror and the distance behind the mirror where your image appears?

The quantitative counterparts to the properties of a plane mirror are expressed as follows:

$$Di = Do \ (\text{Image distance} = \text{object distance})$$

$$Hi = Ho \ (\text{Image height or size} = \text{object height or size}).$$

Experimental Task

(A)   (B)

The diagram shows a garbage can reflected on a plane mirror. If you were the person at (A) would you see the reflection of the can on the mirror?

- Place your finger on the top of the reflection of the lid and mark it on the diagram; if you were to move from (A) to (B), where would you put your finger on the diagram now?

(continued)

- Use a plane mirror supported onto a piece of paper, draw a point (P) in front of the mirror, on the paper.

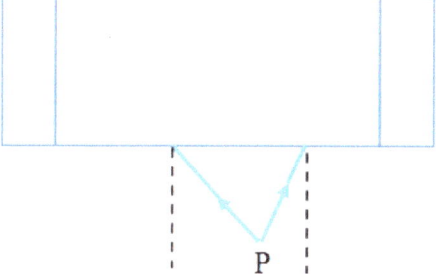

- Draw two rays similar to those shown above on the paper, then draw the normal (the dashed lines) and measure the angles of incidence of both rays. Using the law of reflection, draw the reflected rays, and then their projections until these intersect behind the mirror to form the image of (P).
- As you move your head sideways while looking at the image of (P), does it change the position?

(C)

- If the person moved to position (C), would she still be able to see an image of the can?

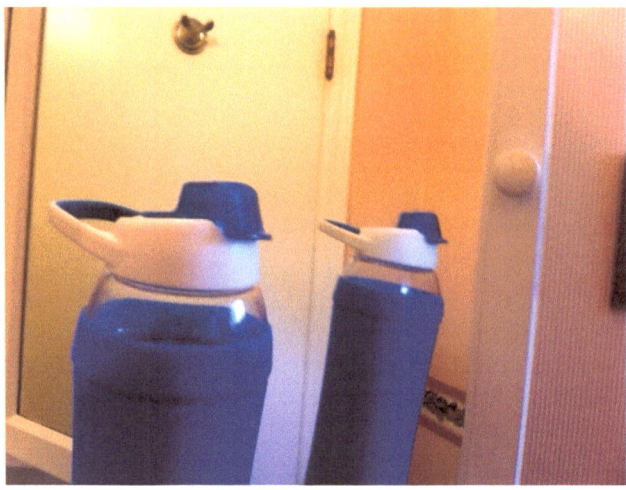

Fig. 3.7 The reflection of the plastic bottle exhibits a well-known but often misunderstood property of plane mirrors. The orientation is unchanged since the handle appears on the left side of both the object and its image; what changes is the front-to-back perspective. The back of the object appears as the front of the image

Fig. 3.8 Placing three plane mirrors at right angles to each other will produce a number of reflections of the single object placed in front of them

Plane mirrors have a unique property that is often misunderstood; as Fig. 3.7 shows, there is a change between the object and its reflection (the image). However, the change isn't in the direction of the orientation, it is in the sense that the front of the image is the back of the object. There are objects and patterns, among them some letters and numbers that will indeed show a reversal in the direction upon reflection from a plane mirror, but that is a property known as symmetry that will be explored later in the book. Nevertheless, the presumed change in orientation attributed to *all* objects reflected from a plane mirror is a misconception arising from experience often lacking in reflection.

A combination of plane mirrors has many interesting applications in the creation of illusions and changes in perspective. Placing mirrors at certain angles gives rise to a variable number of images, depending on the angle between the mirrors, as Fig. 3.8 illustrates.

Activity

A person stands in front of a plane mirror whose size only allows her to see her image from the waist up.

Is there a way that the person can move so that her whole body can be seen on the mirror?

Now try it yourself with a mirror where you can only see the top half your body reflected, and compare your experience with your prediction.

Curved Mirrors

A *concave* mirror is constructed with the silvered surface of the mirror on the inner, or concave, side of the curve. A plane mirror is essentially a special case of a more general type of reflective surface, that of a curved mirror. The difference in perspective is that if one stands very close to a large curved mirror, its curvature may not be apparent and its surface approximates that of a plane mirror. However, there are some definite differences between the images produced by both types of mirrors.

Conceptual Task
Double-sided mirrors used for makeup applications have a very distinctive property when you compare the reflections from both sides, what is it?

The first type of curved mirror we investigate is a concave mirror, as shown in Fig. 3.9.

Figure 3.9 shows that all incident rays in this case are reflected through F (the focal point). This is true for mirrors as well as for satellite dishes, where the antenna will be located at the focal point. Concave mirrors are also called *converging* since all rays coming from far away converge at the focal point upon reflection. We shall regard all values of Do, Di, and f as positive when these locations are in front of the mirror, and negative when they appear behind it. There are potentially an infinite number of rays emanating from all points on the surface of an object that either radiates or reflects light. At this point we shall introduce three rules as shown in Fig. 3.10 that enable us to locate the images formed by mirrors, all rays are assumed to be coming from radiant objects (whether producing or reflecting light):

The dashed lines are the normal lines or radius of curvature of the mirror, showing that the law of reflection is obeyed (the angle of incidence = the angle of reflection).

We can always add any other ray that obeys the law of reflection, provided that we draw the normal to the surface (the radius) so that the angles of incidence and reflection are equal.

As long as we have at least two reflected rays that intersect, it will be enough to locate the image.

We now introduce the equations needed to provide the quantitative description of the properties of mirrors (which will be the same for lenses later).

$$f = \frac{R}{2} \tag{3.1}$$

$$\frac{1}{f} = \frac{1}{Do} + \frac{1}{Di} \tag{3.2}$$

$$M = -\frac{Di}{Do} = \frac{Hi}{Ho} \tag{3.3}$$

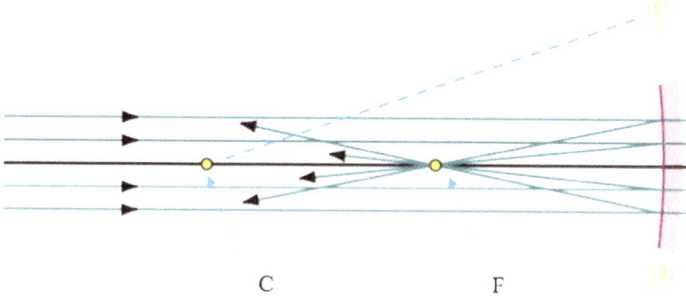

C   F

Fig. 3.9 A concave mirror showing rays coming from an object on the left infinitely far away so that its rays are parallel, and two points; the center of curvature of the mirror (C) where the dashed line, the radius begins, and the focal point (F) halfway between the center and the mirror. The horizontal darker line on which the center and focal point are located is called the principal axis. The distance between the mirror and the focal point (F) is called the focal length (*f*)

Fig. 3.10 Illustration of the three rules for ray reflection. 1. A ray incident parallel to the principal axis is reflected through the focal point. 2. An incident ray through the focal point is reflected parallel to the principal axis. 3. An incident ray through the center of curvature is reflected back through (C)

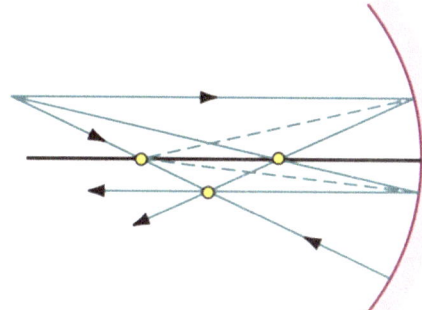

The variables in the equations have all been introduced except for *M* which stands for the lateral magnification, which is taken to be negative when the image is inverted (upside down) as compared to the object. Magnification doesn't mean that it is always larger, it could sometimes be smaller than 1, in which case the image will be smaller/shorter than the object.

Since the equations apply to both plane and curved mirrors, we can see that for a plane mirror $M = 1$ since the image and object distances are the same, as well as the image and object sizes or heights. Note that the only way for *M* to be 1 is for *Di* to be negative, which makes sense since as we saw the image produced by a plane mirror is always behind the mirror. In addition, the radius of curvature for a plane mirror is infinitesimally large, thus with $R = \infty, f = \infty$ and if we substitute this into Eq. (3.2):

$$\frac{1}{\infty} = 0 = \frac{1}{Do} + \frac{1}{Di} = 0 \tag{3.4}$$

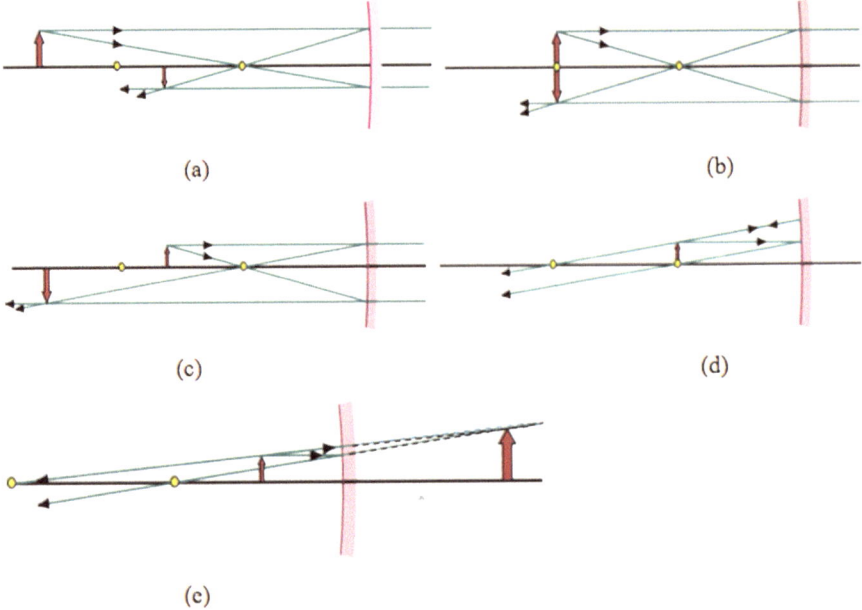

(a)

(b)

(c)

(d)

(e)

Fig. 3.11 Cases where in (**a**) $Do > 2f$, (**b**) $Do = 2f$, (**c**) $2f > Do > f$, (**d**) $Do = f$, (**e**) $Do < f$

Hence $Di = -Do$ (the image distance will always be equal to, but negative as compared to the object distance. That is why the image is considered to be virtual, or appearing to form on the other side of the mirror.

We can now explore the various cases where a concave mirror exhibits interesting properties, all depending on the object's distance, as shown in Fig. 3.11.

In the first three cases an image is produced (the inverted arrow) that is real, but that is smaller in (a), equal in size/height in (b), and larger in (c). There is no image in (d) since the reflected rays are parallel (they never intersect), and the image is virtual, larger, and upright in (d). In the last case the arrow indicates an image but the lines that intersect aren't solid since they are projections, not real rays.

Problem

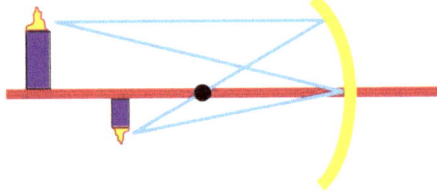

- According to the diagram, what condition of Fig. 3.10 is being represented?
- What two rays are drawn?
- Using Eqs. (3.1) and (3.2), where is the location of the object if the image is located 20 cm in front of a 30 cm radius mirror?

(continued)

Experimental Tasks
(1) Comparing a Plane and a Concave Mirror

Procedure

Place a plane and a concave mirror side by side (makeup mirrors are suitable since one side is concave to magnify features). They don't have to be as large as those shown in the figure, but they should be comparable in size. Make sure you have a large reflective surface opposite to them (such as a blank wall) on which to see the images produced.

- Dim the room lights or turn them off, turn on and place a small flashlight or other bright light source in front of each mirror. Describe what you see as you illuminate each mirror; what is the difference?
- With the lights back on, stand facing both mirrors and describe what you see on each as you move slowly away from them.
- Use the terminology and equations introduced to explain what you have observed.
- Given what you have observed, where is the focal point of the plane mirror?

(2) Using a Concave Mirror
Procedure

1. We first determine the mirror's focal length; this will be done in two ways:

(A) Focus on a distant object (as in Fig. 3.9) by moving the screen until its image appears sharp on it. Record the distance the screen is from the mirror.
Use $\dfrac{1}{f} = \dfrac{1}{Do} + \dfrac{1}{Di}$ where $Do \approx \infty$ (the object is infinitely far away compared to the image distance), to solve for the image distance (Di) which is in this case equal to the focal length (f); Hence $Di = f$.

(continued)

Focal length of mirror =   (I)

(B)   Use a light source as the object and place it at a point so that you can
         find its image on the screen. Record both distances, the object's and the
         screen's from the mirror. Then use

$$\frac{1}{f} = \frac{1}{Do} + \frac{1}{Di}$$

But this time we solve for f since we know both Do and Di
focal Length of mirror =   (II)
You should average the two results (I) and (II)
Average value of mirror's focal length =

A *convex* mirror is constructed with the silvered surface of the mirror on the outer, or
convex, side of the curve. In this case the focal length and the image distance are always
negative, and therefore virtual. Instead of the five cases we saw for a concave mirror, a
convex one exhibits a single case, as illustrated in the figure below. Convex mirrors are
also called *diverging* since the reflected rays spread in different directions.

As Fig. 3.12 shows a convex mirror has an important characteristic that is useful
for many applications. It allows for the creation of images that can be reproduced in
small areas. The objects in the field of view can be large in number, and spread out
over large areas as well.

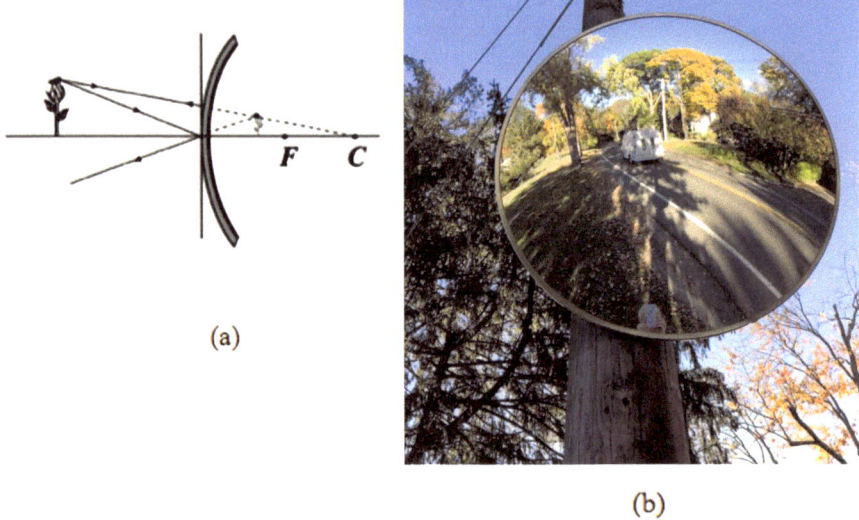

(a)

(b)

Fig. 3.12 A convex mirror always displays the same type of image, regardless of the position of
the object. In (**a**) the two projections of the reflected rays (*the dashed lines*) intersect to produce a
virtual, upright, and smaller image. The most general application of convex mirrors (**b**) is to gather
a large field of view into a much smaller image

Activity

Consider the mirrors used in automobiles for various purposes. There are reasons for two plane mirrors (the driver's side and rear view), and the passenger's side mirror. Why is there a warning on the passenger's side mirror, but not on the others?

The rules for image formation by mirrors have limitations since there are circumstances where images will appear blurred. Parallel incident rays that are far from the principal axis converge to other points besides the focal point, and other light rays make large angles with the principal axis. The result is an effect called **spherical aberration**, where the extended image doesn't form at a single point.

Application to Sound

We now look at the reflection of sound waves, where by contrast very often the more interesting applications are from reflections that occur from irregular or rough surfaces, not from smooth ones. The simplest case is that of the production of an echo, as sound is reflected from a large and fairly smooth surface.

The main application is in the determination of the acoustic properties of performing spaces like concert halls, auditoriums, lecture halls, and even churches and libraries. The property measured is the perceived intensity or loudness of a sound specifically created in the chosen space, and how long it takes for it to decrease until it blends in with other ambient sounds or background noise. This change in loudness is called *reverberation*. The more reflective the surface, the longer the produced sound remains in the air, and so the longer the reverberation time. Highly absorbing surfaces will yield a short reverberation time since the loudness of the sound produced will decay quickly.

The convention is to measure the time that it takes for a sound produced at a certain loudness to decay by 60 dB. This concept (and its units) will be introduced and explained in a later chapter, but it shouldn't prevent one from completing the task; dB are simply convenient units used for sound level measurements. This is often quite difficult to achieve due to the challenge of finding spaces where the ambient sound level is lower than 50 dB. Creating a maximum loudness of approximately 120 dB requires creativity and a powerful device. The ideal spaces tend to be those that are very quiet, although ventilation/heating systems can produce ambient loudness levels that can confound the results. The following figure illustrates a way in which the reverberation time can be determined in a room where the maximum loudness can be around 120 dB. Even if the background noise is significant, all one needs is a decrease of 60 dB, to measure the reverberation time; in this case it is about 3 s (Fig. 3.13).

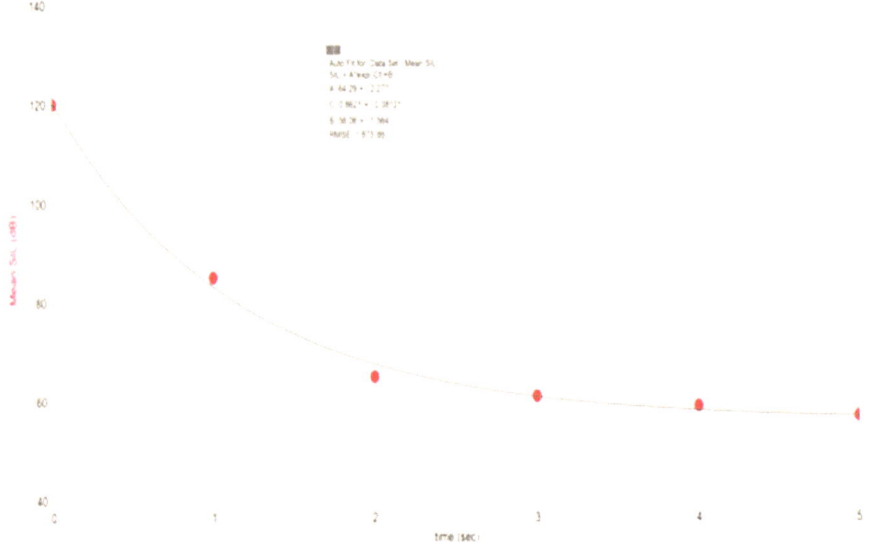

Fig. 3.13 Measurement of the sound changes in a room where readings were taken with a Sound Level Meter. Notice that the reverberation time is approximately 3 s (the time for the sound intensity to decay by 60 dB)

The experimental determination of the reverberation time can be compared to a formula that takes into account the volume of the room, and the types of materials making up all surfaces upon which the sound reflects. The optimum reverberation time for an auditorium or room of course depends upon its intended use. For music and speech reproduction rooms the range is typically between 1.5 and 2.5 s. A classroom should be much shorter, less than a second, and a recording studio should minimize reverberation time in most cases for clarity of recording.

The reverberation time is strongly influenced by the absorption properties, known as coefficients of the surfaces enclosing a room. The larger the room, the longer one would expect the sound to remain in it if the surfaces are not particularly rough or irregular; as a rule one should not expect to get a long reverberation time with a small room.

Chapter 4
Refraction

Conceptual Task

If you put a pencil half way inside a glass filled with water, what do you notice?

Virtual Experiment

The activity can be accessed at https://phet.colorado.edu/sims/html/bending-light/latest/bending-light_en.html.

Make sure the screen looks like the figure below

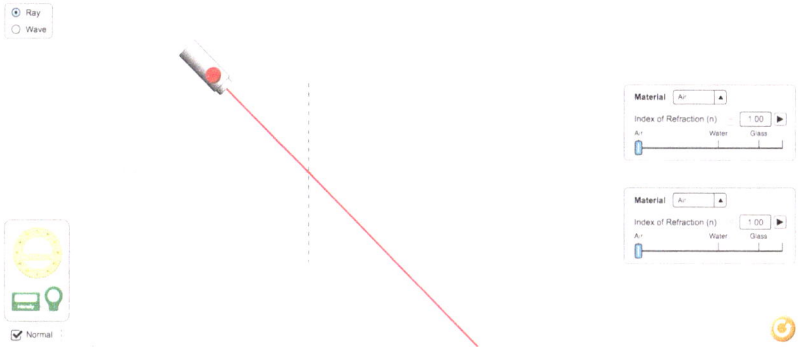

Change the material in the bottom half of the figure to water and then to glass, describe what you see in each case.

© Springer International Publishing Switzerland 2017   75
F. Espinoza, *Wave Motion as Inquiry*, DOI 10.1007/978-3-319-45758-1_4

What is Refraction?

Refraction is the bending of waves as they go from one material or medium into another. As we have previously stated, when waves travelling through a transparent medium encounter a boundary between that medium and another, they are generally partly reflected and partly transmitted or absorbed. When waves travel from one medium into another, their amplitude, wavelength, and speed will change, but not their frequency. The direction of the bending depends on the difference between the materials. Figure 4.1 shows the changes in direction of waves as they are refracted.

Task

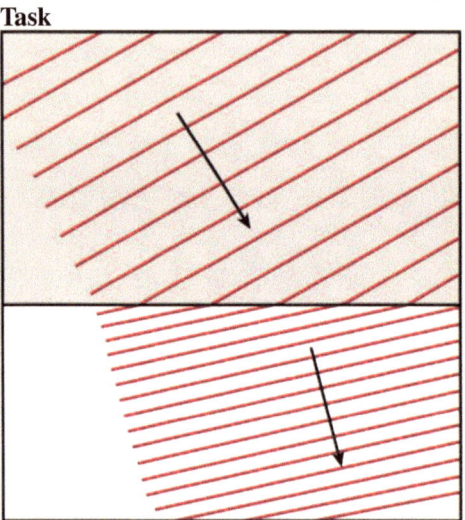

In the diagram the spacing between wave fronts can be used to describe the wavelength; using the formula $v = \lambda f$ and remembering that upon entering a new medium the frequency of a wave doesn't change, what can one say about the speed of the wave in the medium (the lower part) where the bending occurred?

We begin by defining a property of materials known as the index of refraction. It is the ratio of the speed of light through a vacuum to its speed through the material, as stated by Eq. (4.1).

$$n = \frac{c}{v} \tag{4.1}$$

Fig. 4.1 Representation of wave fronts and corresponding rays as a wave enters a different medium than the one through which it has been initially travelling; the change in the direction indicates a more refractive medium than the original one

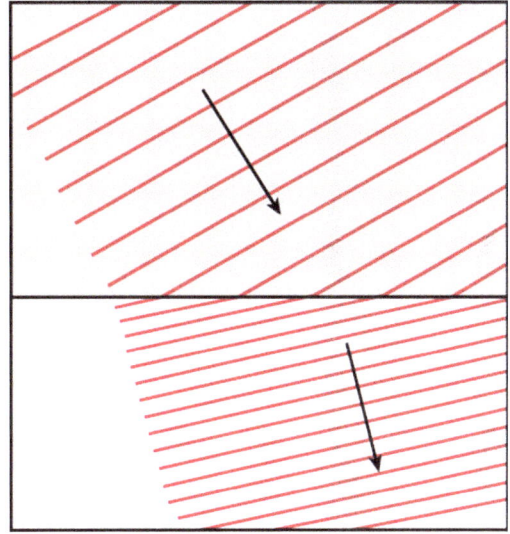

As a reference we take the index of refraction of air to be $n = 1$, which states that the speed of light is the same through a vacuum as it is through the air. This is of course an approximation, since the speed of light is taken to be 300,000 km/s through a vacuum, and it is known to slow down by about 70 km every second as it travels through air. However, 70 compared to 300,000 is insignificant for our purposes, and so we take both speeds to be the same.

As a general rule we state that when light enters a medium of higher index of refraction, the rays bend *toward* the normal, and when entering a medium of lower index they bend *away* from the normal. How much is the bending? It can be determined by the use of Snell's law, after Willebrord Snell who proposed such description in the early part of the seventeenth century.

$$n_1 \sin\theta_1 = n_2 \sin\theta_2 \tag{4.2}$$

In Eq. (4.2) n_1 and θ_1 are the index of refraction and angle of incidence in the first medium; n_2 and θ_2 are those in the second medium.

Examples

1. A light ray travelling through air strikes a glass surface at an angle of 40°, what is the angle of refraction of the ray in the glass?

For air $n = 1$, for glass $n = 1.5$.

Fig. 4.2 Illustration of a
case where light goes from
air into glass

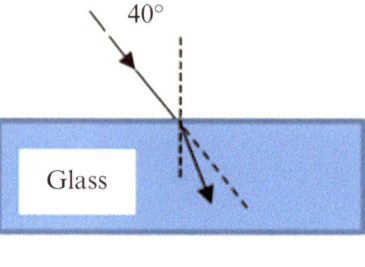

Fig. 4.3 Illustration of the
reverse case where light
goes from glass into air

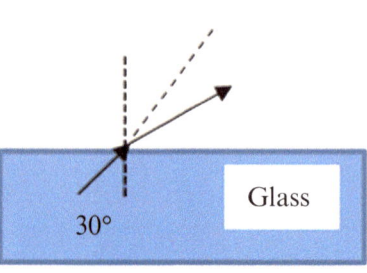

Figure 4.2 shows the incident ray and angle of incidence, measured from the
normal; it also shows that the ray does not continue straight since it enters the glass,
which has a higher index of refraction.

Using Snell's law:

$$n_1 \sin \theta_1 = n_2 \sin \theta_2$$

$$(1)\sin 40° = (1.5)\sin \theta_2$$

$$\sin \theta_2 = \frac{(1)\sin 40°}{1.5} = \frac{0.643}{1.5} = 0.429 \quad \text{therefore,} \quad \theta_2 = 25.4°$$

The angle that the arrow makes with the normal in the glass is 25.4°.

2. Suppose now that a light ray has been travelling through glass and it strikes a
glass–air interface at an angle of 30°. What is the angle at which the ray emerges in
the air?

Figure 4.3 shows the situation; in this case

$$n_1 \sin \theta_1 = n_2 \sin \theta_2$$

$$(1.5)\sin 30° = (1)\sin \theta_2$$

$$\sin \theta_2 = \frac{(1.5)\sin 30°}{(1)} = \frac{0.75}{1} = 0.75 \quad \text{therefore,} \quad \theta_2 = 48.6°$$

The angle that the arrow makes with the normal in air is 48.6°, and that is why the
ray bends away from the normal instead of emerging as the dashed line indicates.

3. If instead of glass a light ray strikes an air–water interface at an angle of 40°, what is the angle of refraction in water ($n = 1.33$)?

Activity

We see objects under water as long as there is light reflected from them to reach our eyes. However, there is a distortion of the way they look. Using Fig. 4.4 explain how the reflected ray that the person sees (2) appears to be coming from the position where the fish seems to be (B), rather than from where it really is (A). Use the result from Example 3, along with the letters and numbers shown, as part of your description.

Total Internal Reflection

There is a special case when light that has been travelling through a medium encounters an interface between the medium and another, the latter one having a smaller index of refraction. Upon striking the interface, the angle of incidence will have an upper limit beyond which something interesting happens.

Figure 4.5 shows the situation in detail. In part (a) both rays (1) and (2) will emerge refracted away from the normal at the glass–air interface. For ray (3), Snell's law states that

$$n_1 \sin\theta_1 = n_2 \sin\theta_2$$

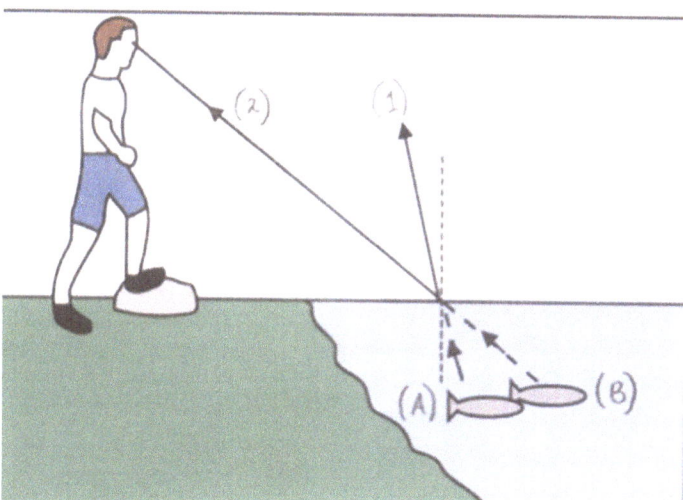

Fig. 4.4 Illustration of a person looking into water and seeing a submerged object (the fish) at a depth different from where it really is

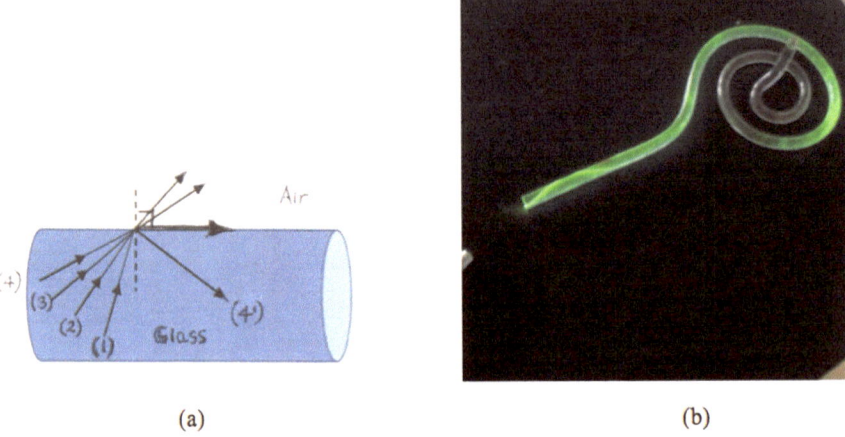

(a)   (b)

Fig. 4.5 Demonstration of total internal reflection. In (**a**) rays (1) and (2) are incident on the glass–air interface. Since glass has a higher index of refraction than air, the refracted rays will bend away from the normal, as shown by the *two top arrows* in air. For ray (3) the refracted angle will be 90°, and it is shown as the *thicker arrow* drawn parallel to the glass–air interface. Ray (4) exceeds the critical angle and so it is reflected (4'). In (**b**) we can see that the laser light travels along the plastic fiber since the initial direction of the light as it enters the fiber is such that the angle of incidence at every instance will exceed the critical angle

As shown in the figure, n_1 is the index of refraction of glass, and θ_2 will be 90°. Since sine of 90° is 1, and $n_2 = 1$, then θ_2 becomes the critical angle θ_c. Therefore,

$\text{Sin } \theta_c = \dfrac{1}{n_1}$, which means that in general, when light has been travelling through a medium and upon striking an interface with another medium of lower index of refraction (in this case air), the angle of incidence becomes the critical angle when the refracted ray travels along the interface. Any higher angle of incidence will result in the rays no longer refracting but being internally reflected. The successive angles of incidence will result in the ray travelling down the medium. Figure 4.5b shows the most obvious application, that in fiber optics for telecommunications.

Exercise

What material has a higher critical angle, water or glass?

Use $n_{\text{water}} = 1.33$ and $n_{\text{glass}} = 1.52$

For water: $\text{Sin } \theta_c = \dfrac{1}{n} = \dfrac{1}{1.33} = 0.751$, and so $\theta_c = 48.7°$

(Repeat for glass and then compare the angles).

Exploratory Task

In the accompanying figure a battery is shown refracted by a glass prism.

1. Why are there two images produced?
2. Draw rays that can be used to construct the formation of those images.

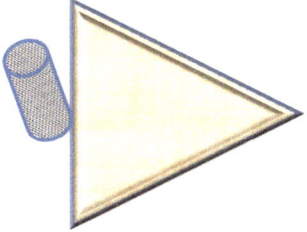

Properties of Lenses

We now turn our attention to the investigation of the properties of lenses, and their applications. As in the case with mirrors, we make a distinction between converging and diverging lenses; except that the types are reversed. In other words, the images produced are the same for converging mirrors and lenses, but a converging mirror is concave, whereas the corresponding lens is convex. Similarly, the image is the same for a diverging mirror, as it is for a diverging lens, but a diverging mirror is convex, whereas a diverging lens is concave. This is due to the curvatures of the surfaces upon which the rays are incident from the objects. In addition, a lens has two focal points since there are two surfaces of specific curvature that contribute to the image formation.

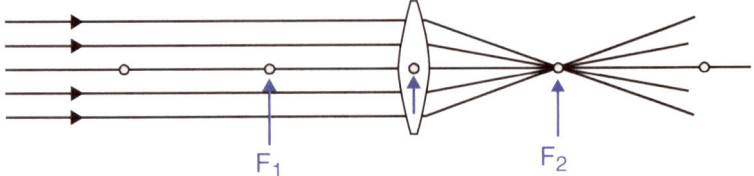

Fig. 4.6 Representation of the convergence of rays to form an image by a convex/converging lens. The *arrows* indicate the location of the first and second focal points, as well as the optical center (which replaces the center of curvature for a mirror)

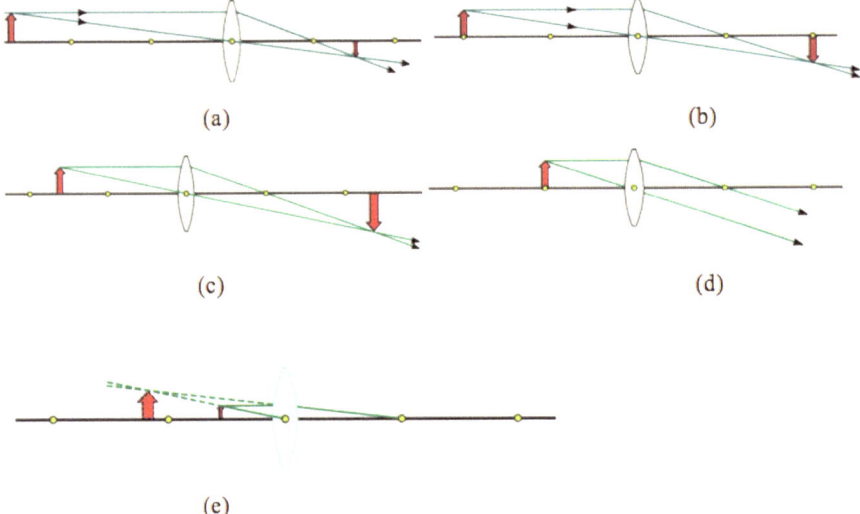

Fig. 4.7 Cases where in (**a**) $Do > 2f$, (**b**) $Do = 2f$, (**c**) $2f > Do > f$, (**d**) $Do = f$, (**e**) $Do < f$. The thickness *of the arrow* representing the image varies as the length does, which happens for (a), (b), and (c)

Figure 5.6 corresponds to the one used in the previous chapter to represent a concave/converging mirror, where all rays are coming from an object far away so that they are parallel, and thus converge at the focal point. In the case of a *convex/converging* lens, the parallel rays are refracted through the second focal point (the one at the far side of the lens) (Fig. 4.6).

The same formulas apply as with mirrors, but the criteria for rays are slightly changed:

1. An incident ray parallel to the principal axis is refracted through the second focal point (F_2).
2. An incident ray through the first focal point is refracted parallel to the principal axis.
3. An incident ray through the optical center (instead of the center of curvature) does not undergo refraction; in other words, it goes straight through the lens without changing direction.

In the first three cases shown in Fig. 4.7 an image is produced (the inverted arrow) that is real, but that is smaller in (a), equal in size/height in (b), and larger in (c). There is no image in (d) since the refracted rays are parallel (they never intersect), and the image is virtual, larger, and upright in (d). In the last case the arrow indicates an image but the lines that intersect aren't solid since they are projections, not real rays.

Figure 4.8 is a detailed illustration of the last case (e) in Fig. 4.7 in the case where an object is located closer to the lens than the focal point.

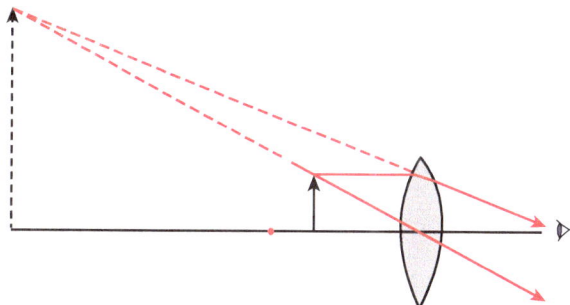

Fig. 4.8 Fig. 4.7(e) is worth showing in more detail, as it represents a very important application of the use of a convex lens, that of a magnifying glass. Note that the eye is shown at the location where one would have to be, to see the image (*the larger arrow*) produced by the object (*the short arrow*) when it is closer to the mirror than one of its focal points (F_1)

Exercise

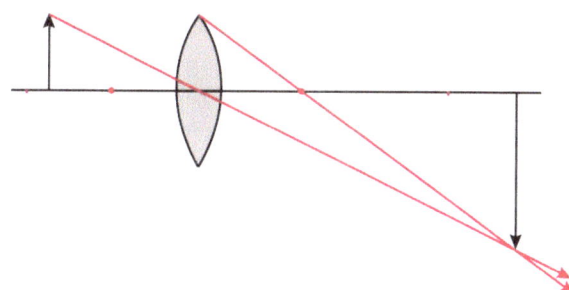

(A) In the accompanying figure, draw the missing ray coming from the object and refracting in such a way as to intersect the other ray and form the image shown.
(B) Which of the three rules does that ray follow?
(C) Use the lens equation $\dfrac{1}{f} = \dfrac{1}{Do} = \dfrac{1}{Di}$

to find the image distance (Di), if the object is located at 18 cm (Do), and the focal points are at 10 cm (f)?
(D) Does the numerical answer confirm the location of the image in the diagram?

(continued)

Exploratory Task

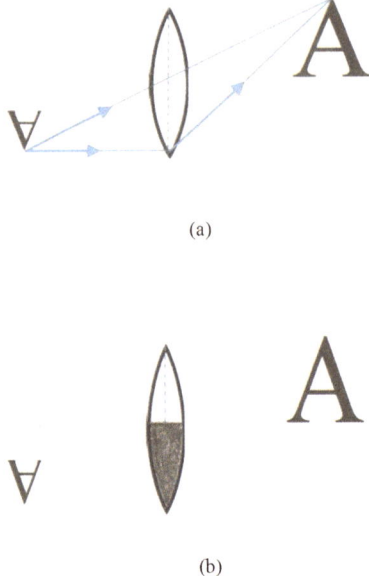

(a)

(b)

Part (a) shows an object (the inverted letter A) located such that $2f > Do > f$; two rays originate at the vertex and intersect at a corresponding point to produce the image. In part (b) the lens is half-covered with a dark material.

1. What two rules are followed by the rays in part (a)?
2. Is it possible for an image to form in part (b)?
3. If you think an image can be formed, provide evidence by drawing any two rays that can intersect to form it.

A *Concave*/diverging lens is constructed with the curvature of the refracting surfaces reversed from that of a *convex* lens.

Figure 4.9 shows that the image produced by a concave lens is like that produced by a convex mirror. It is important to notice that when applying the rules for the refracted rays, the one incident parallel to the principal axis is now refracted through F_2 instead of F_1 as with a convex lens. In both cases the image is perceived to be located on the same side of the lens/mirror as the object, and it has the same characteristics regardless of the position of the object. All three rays are drawn in Fig. 4.9a to show how the image forms.

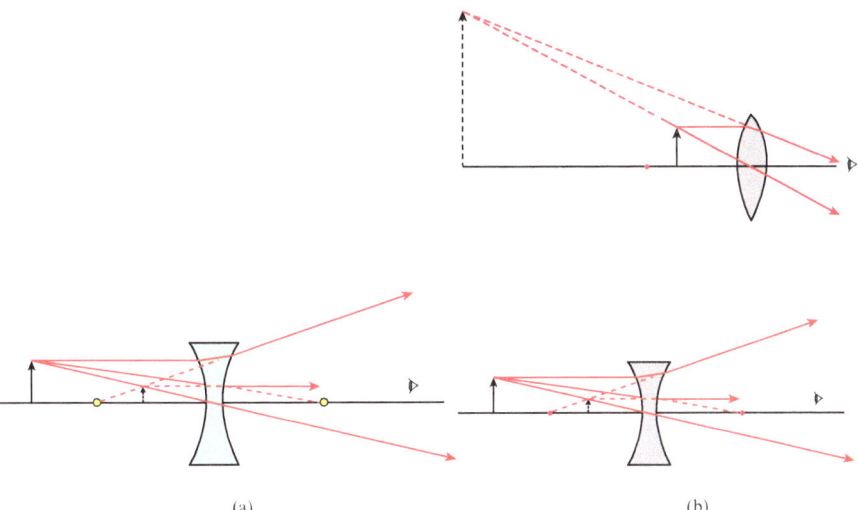

(a)   (b)

Fig. 4.9 Image produced by a concave lens (**a**). A concave lens produces an image that like that of a convex mirror is always smaller than the object, upright, and virtual. The comparison between the virtual images produced by a convex and a concave lens is shown in (**b**). In both cases the eye perceives a virtual image even though it cannot be projected onto a screen; the difference is in the size of the image

Exercise

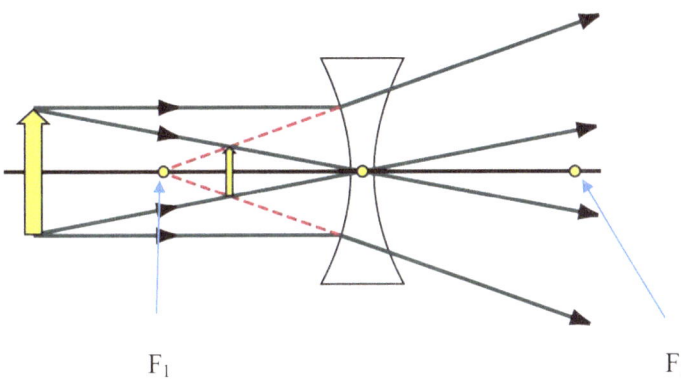

F_1 F_2

In the above diagram

(a) What two rays are shown to form the image?
(b) Why are the incident rays parallel to the principal axis refracted through F_1 instead of through F_2 as was the case with a convex lens?
(c) Use the lens equation $\dfrac{1}{f} = \dfrac{1}{Do} = \dfrac{1}{Di}$ to determine the location of the image (Di) if the lens has a focal length of 20 cm, and the object is located at 80 cm.
(d) Is there something contradictory between your answer to c) and the location of the image in the diagram?

As pointed out in Chap. 3 parallel incident rays that are far from the principal axis converge to other points besides the focal point, and other light rays make large angles with the principal axis, thus producing a blurred image. A similar situation occurs with lenses, where the result is an effect called **chromatic aberration**; again, the extended image doesn't form at a single point. This last observation leads directly to a very important application of lenses, where vision problems arise due to conditions related to these so-called aberrations.

Applications of Lenses

The human eye has a structure that contains parts that behave in a similar way to mirrors and lenses. Vision problems have been corrected for centuries with the use of lenses, and modern interventions such as laser surgery continue to improve the treatment of vision afflictions.

Figure 4.10 shows a diagram of the basic structure of the human eye. The main parts of the eye that concern us are the cornea, the aqueous humor, the lens, the retina, and the optic nerve. As the diagram indicates, the cornea, the aqueous humor, and the lens appear to have properties similar to those of lenses. Vision issues related to these parts can be corrected according to the needed refraction properties associated with lenses, whether the corrections are accomplished with eyeglasses, contact lenses, or laser surgery. In such cases the curvature and index of refraction of the visual aids can be manipulated accordingly.

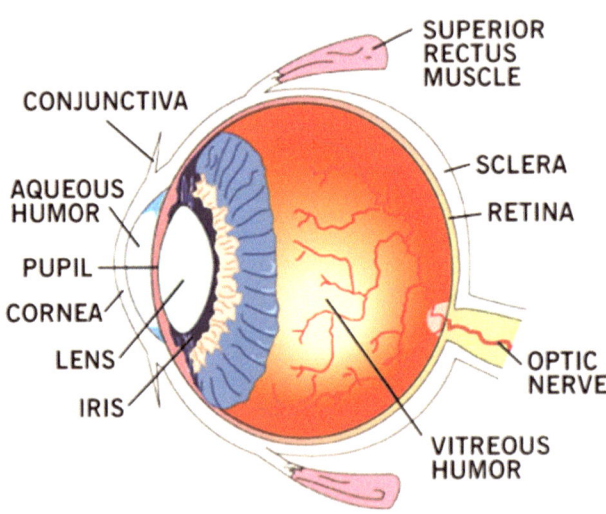

Fig. 4.10 Illustration of the basic structure of the human eye

The two most common vision problems involving the eye's ability to focus on objects can be corrected with convex and concave lenses, as well as with changes in the index of refraction. *Myopia* or near-sightedness is the inability to focus on distant objects results from images forming in front of the retina, and is due to the index/curvature of those parts that contribute to image formation. In this case the afflicted person needs to move distant objects closer to the face to be able to focus on them. *Hyperopia* or far-sightedness is the opposite case, where the image forms behind the retina and so the person needs to move near or proximate objects farther away to be able to focus on them.

Conceptual Task

Knowing the reasons for both Myopia and Hyperopia, using the lens of the eye draw two rays coming from the object and forming an image in front of the retina in (a), and behind the retina in (b).

What kind of lens can correct each of them? Explain.

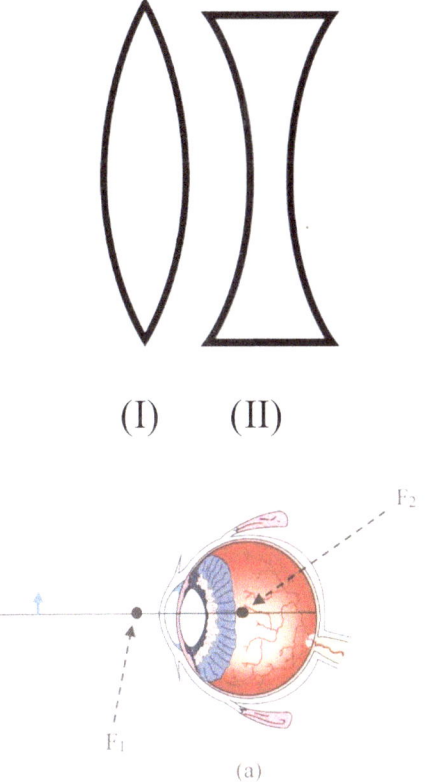

(I)   (II)

(a)

(continued)

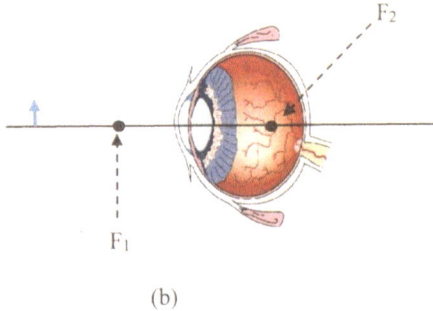

(b)

The next activity is based entirely on the role played by the retina and the optic nerve, in terms of the reflective properties associated with mirrors.

Experimental Activity
Blind Spot Properties
  Background
  Each eye has a surprisingly large blind area, which is about 4° of the visual angle, roughly the width across your four fingers held at arm's length. Fortunately for us, they are in different locations in each eye, the one in the left eye is about 10° (two hand widths at arm's length) to the left of the central visual region, and the one in the right eye, an equivalent distance on the other side. Amazingly, we are normally unaware of these natural blind spots. They are either filled in perceptually (a remarkable phenomenon) or they are ignored and so not seen. These are very different possibilities for explaining why the eyes' blind spots are not generally seen or noticed, even when one eye is covered.
  Examples of effects related to the blind spot

(continued)

To get an idea of what the blind spot looks like, and what it does, look at each case above *with your right eye closed*. (1) Tilt the page until the + becomes an x; as you move your head away/towards the page, what do you notice happening to the top figure? (2) Keep the page vertical (without any tilt) and repeat; describe what you see in the lower figure. In this activity we explore two features of the blind spot:

(1) Variation of distance where the spot disappears from the surface (wall) as a function of distance between the spot and a reference (+) position. Is it symmetrical for both eyes? Suggestion: two sets of paper/index cards should be made, one with (+) on the left (as shown), and one with (+) on the right (for the left eye, with the right one closed).

(2) Variation of blind spot size (diameter) measured between the two points where the spot disappears and then reappears, as a function of distance from the surface where this takes place (wall).

(1) **Reference**

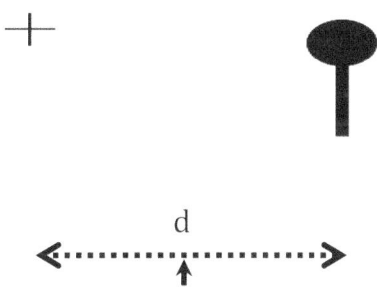

(continued)

Distance between spot and reference d (cm)	Distance from surface where spot disappears D (cm)	
	Left Eye	**Right Eye**

(continued)

(2)

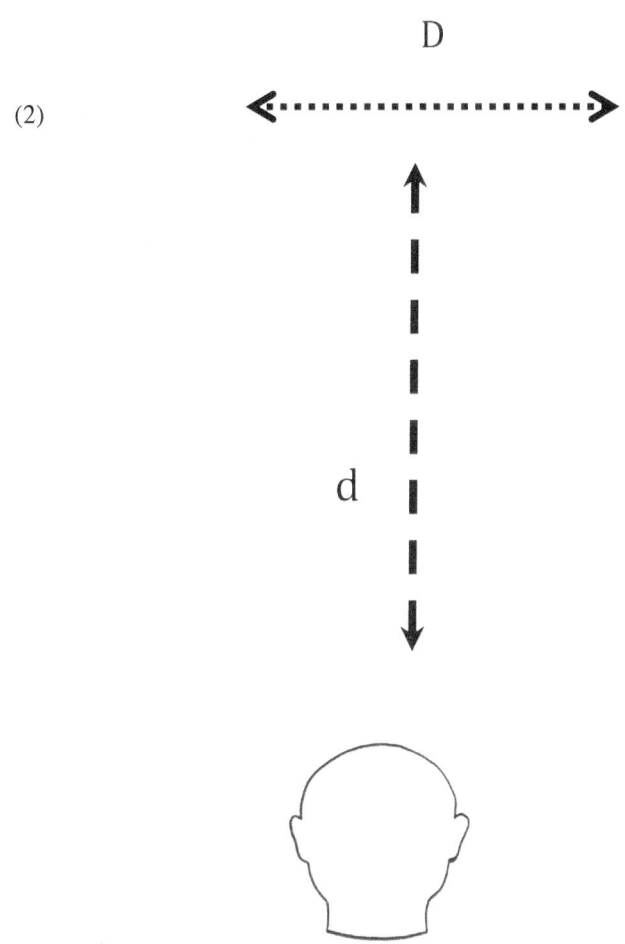

(continued)

Distance from surface (wall) d (cm)	Size of blind spot D (cm)

Plot the data from the tables, D (vertical) vs d (horizontal) in both cases.

Note: Only one graph is needed from the first part, either the left or the right eye data.

Examples of Data Graphs

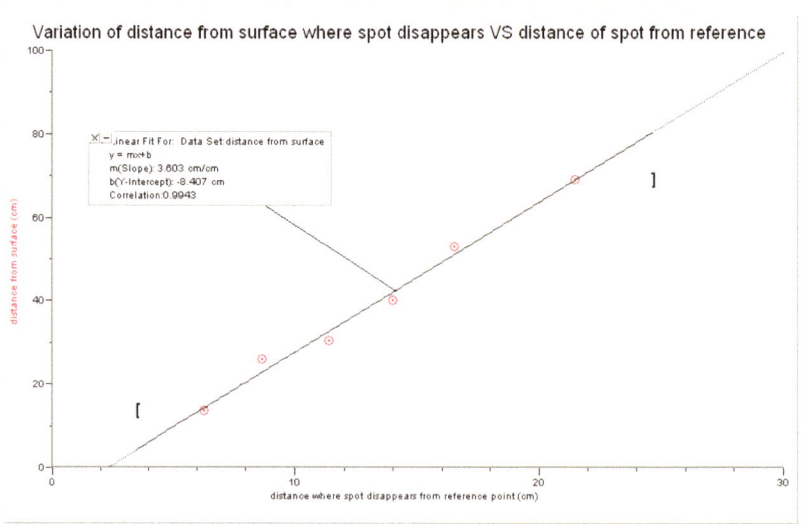

Graph of a sample data collection for the first table

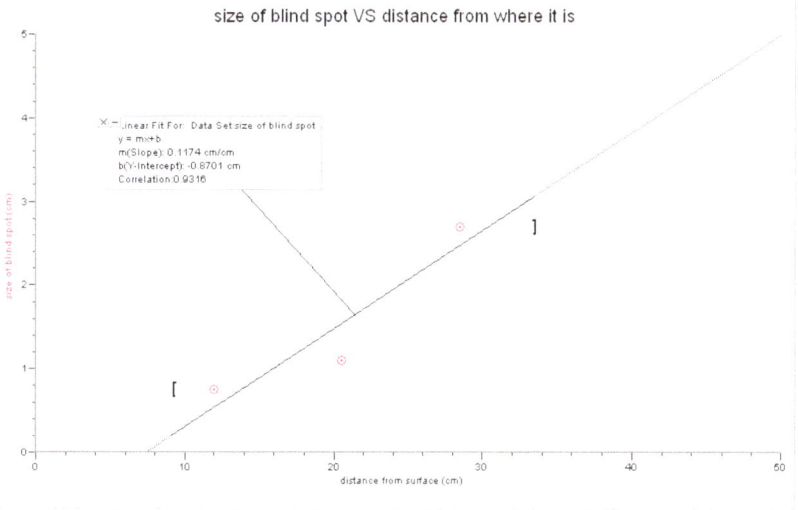

Graph of a sample data collection for the second table
Reflections

- Do the data look symmetrical (roughly the same distances) for the left and right eye in the first table? Explain.
- Compare your graphs with the ones included, and discuss the sources of error in this experiment.
- Consider the situation of an animal that has only monocular vision (one eye on each side of the head) in terms of the second graph. What does the relationship found between the size of the blind spot and the distance to where it is suggest that such animals must be attentive to?
- How can you relate your conclusion to the animal's behavior, particularly when they are eating?

Experimental Task
Snell's Law, Refraction, and Total Internal Reflection
We explore the properties of refraction and use these to determine the indices of refraction for glass and water, as well as the critical angle for which total internal reflection occurs.

(continued)

Part I. Refraction

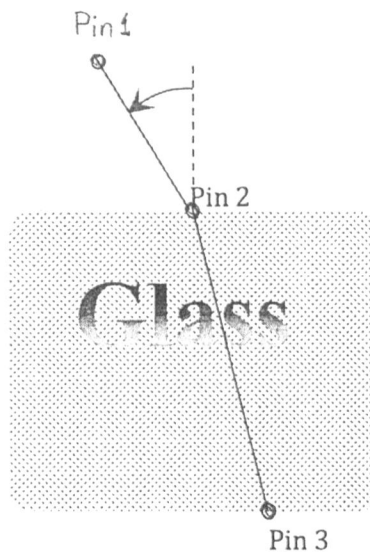

Place a flat piece of glass onto a sheet of white paper so that it looks like the figure above.

1. Place two pins to form a ray and view them *by looking through the glass* so that they appear as one, this way you can create a ray that is refracted (shifted) as you view the pins through the glass, and compare the view with that through air. Put a third pin at the point where you see the refracted ray emerging from the glass, on the other side of where the pins are initially placed.

2. Connect the line that makes the incident ray at the point it hits the glass, to the third pin on the other face, so that you can reconstruct the ray as it travels through the glass.

3. Measure the angle of refraction (from the normal to the ray made by the line connecting pins 2 and 3).

(continued)

4. Fill in the table below as you change the angle of incidence.

Angle of Incidence (θi)	Angle of refraction (θr)	Sin θi	Sin θr
10°			
20°			
30°			
40°			
50°			
60°			

5. Plot the data so that the values for Sin θr are along the x-axis (the horizontal), and those of Sin θi are along the vertical.

6. Determine the slope of the line formed by the points.

7. Repeat the procedure using the semicircular container shown below. The difference now is that the container is filled with water.

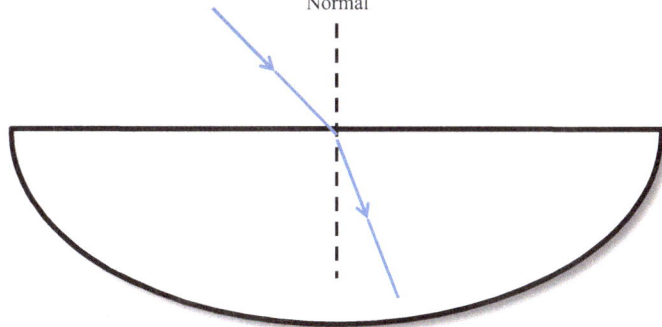

Normal

Place the container onto polar graph paper so that the angles can easily be read off itit; otherwise, a protractor must be used to measure angles if plane paper is used. Outline the rays as you did with the glass plate by using the pins. Vary the angle of incidence as before and record the data on the following table:

(continued)

Angle of Incidence (θi)	Angle of refraction (θr)	Sin θi	Sin θr
10°			
20°			
30°			
40°			
50°			
60°			

8. Upon determining the slopes of both graphs (the one constructed for glass and that for water), compare their values with the indices of refraction of each substance, respectively.

Part II. Total Internal Reflection

1. For this part we use a laser (be very careful not to point it at anyone's face!). Using the same semicircular container, place it upside down so that the incident laser ray strikes the curved surface, and upon entering the water does not bend since it will always strike the container along the normal (the radius of the circular surface).

2. Vary the angle of incidence until the transmitted ray is reflected at the flat surface. When that happens, record the value of that angle to determine the critical angle for total internal reflection in water. The situation is depicted

(continued)

in the figure below, (a) shows the way the rays enter on the curved surface, and (b) what the reflection should look like at the bottom face when the angle is the critical angle.

(a)

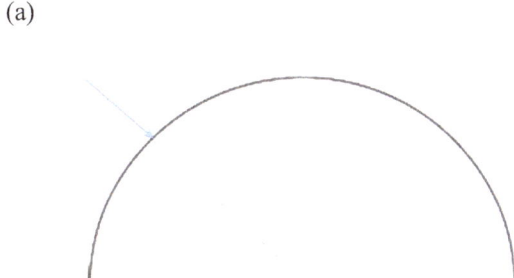

(b)

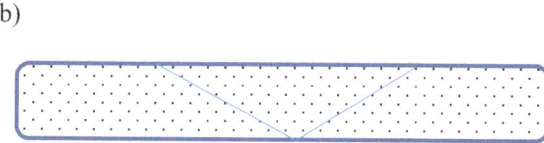

3. Determine the experimental critical angle twice and find its average, compare the experimentally determined critical angle to that predicted by theory using

$$\sin \theta c = \frac{1}{n} \rightarrow \theta c = \sin^{-1} \left(\frac{1}{n}\right)$$ where n is the index of refraction of water (1.33).

4. Determine the percent error by using

$$\% \, error = \left| \frac{\theta c \, Th. - \theta c \, exp.}{\theta c Th.} \right| \times 100$$

The % error is the absolute value of the difference between the theoretical value θc above using n = 1.33, and the experimental value of the measured angle. This difference is then divided by the theoretical value, and then multiplied by 100.

Critical angle	Critical angle	% error
(θc exp)	(θcTh)	

Reflections

Application to Sound

The refractive properties of sound waves that we shall consider have applications in medicine as diagnostic tools, in echolocation, and in outdoor activities, as sound propagates through media where its speed exhibits variable properties. Figure 4.11 shows a situation that may be familiar to you. It is included to encourage you to explore an application.

Figure 4.12 serves as an explanation for Fig. 4.11 and illustrates the case where sound is refracted as it travels through the air, due to the change in direction resulting from the fact that the speed of sound depends on the temperature of the air. The left side illustrates what happens as a source on the surface produces sound and it propagates through the air. A listener (a) at the same level as the source will hear the direct sound represented by the arrows closer to the ground on either side. (b) indicates that the listener hears additional sound produced by the contributions of the higher layers of air, whenever there is a temperature inversion.

Normally the surface would be radiating heat and as the height above it increases, the temperature decreases. However, during temperature inversions, such as when the air over a lake in the early morning is being heated by the sun, the reverse will take place; namely, the temperature will be lower over the surface of the water than in layers above it. The result is that sound will propagate faster in the warmer layers than in the cooler ones, and so when sound emanates from a source there will be a contribution from the speed at higher temperatures, to what the direct value is along the surface of the water. Consequently, in such a case someone will hear sounds from far away that would not be audible at other times.

Fig. 4.11 Does this situation look familiar to you? What do you suppose it represents?

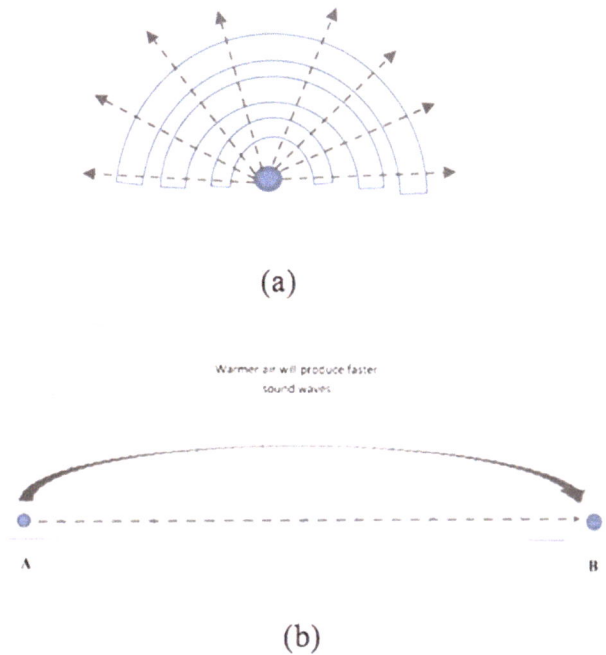

(a)

(b)

Fig. 4.12 Emission of sound waves from a source; they are represented both by circular wave fronts and arrows perpendicular to them. In (**a**) a listener at the same height as the source on the ground will received the direct sound produced by the two lowest *dashed arrows* (one on each side). The other *arrows* represent the sound as it propagates through the air. In (**b**) due to temperature inversion caused by the sun heating the air in the early morning resulting in cooler layers near the surface, the wave fronts bend to indicate the refraction of sound. The net result is that a listener at B will experience sound coming from the direct (*dashed arrow*) direction, as well as from the *bent arrow* at higher layers

Exercise

The dependence of the speed of sound on temperature is given by the equation

$$v = \left(331 + 0.606 T_c\right) \text{m} / \text{s}$$

where T_c is measured in degrees Celsius.

In each of the cases represented below, the person at A is at the source of the sound so she does not measure a value for the speed of sound; however, the person at B measures the value given in each case.

(continued)

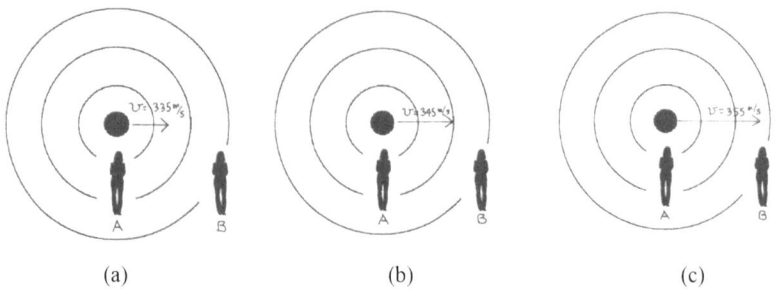

(a)   (b)   (c)

Using the above equation, what must be the air temperature for (a), (b), and (c)?

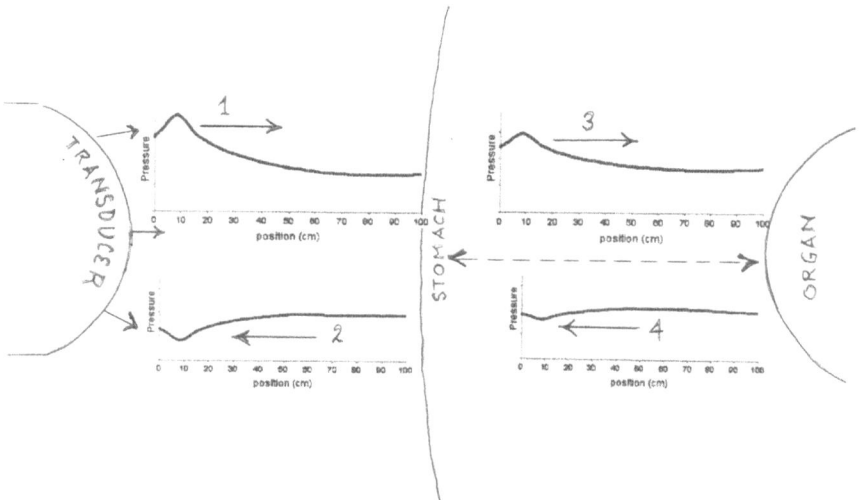

Fig. 4.13 Use of sonography for diagnostic and treatment purposes. The basis of the technique is a sound pulse sent by the transducer (a device that both sends and receives signals) that is both reflected and transmitted. Upon encountering the stomach wall, pulse 1 is reflected (pulse 2) and transmitted (pulse 3). The orientation is reversed for all reflected pulses shown since the incident pulses (1 and 3) encounter a boundary with a medium of higher index of refraction. The distance to the organ (*dashed black arrow*) can be determined by using the time difference between pulses 2 and 4, and using the speed of sound inside the body

In the case of sonography or the use of sound in medical applications, high frequency (ultrasound) pulses are used and their echoes or reflections help determine the locations and other features of internal organs as the sound encounters objects of varying densities in the body.

Consider a pulse incident upon a person's body and experiencing both reflection and refraction at two obstacles; the first is the stomach surface, and the second an internal organ. The following figure illustrates the situation.

Figure 4.13 illustrates the basic procedure where an incident sound pulse undergoes reflection and transmission three times; twice at the stomach wall and once at the organ surface. The last reflected pulse (pulse 2) is really an approximation of the transmitted one from pulse 4; at the stomach wall pulse 4 is both reflected and transmitted. However, the reflected part is minimal since the sound is now going from a medium of higher (the body) to one with a lower index of refraction (air), thus the transmitted pulse (pulse 2) contains most of the sound energy. These details were left out of the figure to keep it as simple as possible.

Problems

1. Ultrasound waves are used for imaging and treatment; if the speed of sound inside the human body is approximately that of sound through water (1500 m/s), and if the range of imaging frequencies is between 2 and 15 MHz.

   (a) What is the range of wavelengths visible?
   (b) What are the types of everyday objects that are visible with this technique?
   (c) If 3.5 MHz is used for abdominal imaging, what is the smallest size visible?

2. If your dog accidentally swallows a pebble roughly 5 mm in size, can ultrasound imaging be used to determine its location? Explain why or why not.

Chapter 5
Interference and Standing Waves

Have you noticed that in a room full of people you are able to hear several conversations, and it is often difficult to concentrate on the one you may be having with the person next to you? Or that if an object falls into a liquid it creates ripples that mix with one another after a while?

This is the result of the fact that waves behave very differently from other material objects upon encountering other waves, or even encountering themselves, which sounds rather odd. In general, whenever an object interacts with another, such as when they collide the result is either a separation or a joining of them. This is a consequence of the properties possessed by the objects known as energy and momentum that are exchanged between the objects regardless of whether they separate or stay connected after the collision.

Waves, on the other hand, exhibit the ability or capacity to go through one another as they interact, resulting in properties different from other objects. This statement of course is only true at the level of perception where the laws of everyday phenomena apply. There is an area of physics known as quantum mechanics where both material objects such as sub-atomic particles, and waves behave similarly. This effect dissipates as the size of the objects increases beyond the realm of microscopic interactions. Therefore, the terminology we use is strictly applicable to phenomena where quantum effects are negligible.

The Principle of Superposition

As pointed out in chapter two when discussing pulses, whenever we use the term interference as applied to the way material objects or people interact, we mean an obstruction of sorts, where something stands in the way, or impedes something else. In the case of waves interference means the opposite; the interaction is not repulsive but cohesive if you wish to look at it that way. The interaction of waves can be understood with the principle of superposition. It basically states that waves can

© Springer International Publishing Switzerland 2017   103
F. Espinoza, *Wave Motion as Inquiry*, DOI 10.1007/978-3-319-45758-1_5

be combined at the same location in space, and this combination leads to two types of interference, constructive and destructive.

In constructive interference the resulting amplitude is greater than the individual ones before they interacted, and the opposite is the case for destructive interference, where the waves can even cancel each other out in terms of amplitude. However, the individual waves can pass through each other without being altered in any way, the alteration takes place afterwards. If the shape of the waves is not maintained, the resulting wave will eventually dissipate, as energy is lost in the interaction. However, if the shape is kept, such as when a string is held fixed at both ends, and is continually plucked, then an interesting phenomenon is produced, that of a *standing wave*. The following figure illustrates this for the case of mechanical waves produced by a vibrating string held fixed at the ends. What is interesting is that this represents the case when a single wave can interfere with itself to produce both constructive and destructive interference.

Figure 5.1 illustrates that as the wave produced by the string bounces back and forth between the ends, a repetitive pattern results where constructive interference is a maximum at the largest amplitudes, and destructive interference a minimum at the points along the horizontal where there is zero amplitude. These alternating points are known as both the antinodes (A) and the nodes (N). The figure shows the first four modes or patterns of vibration—the fundamental (bottom) and its multiples or harmonics. Note that the fundamental contains two nodes and one antinode.

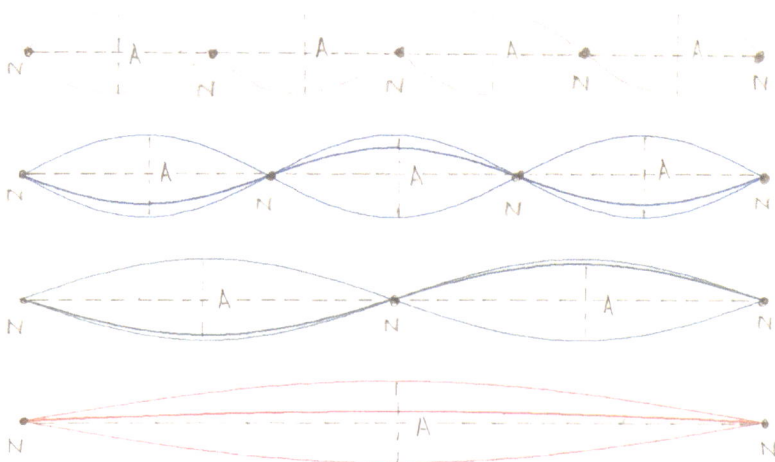

Fig. 5.1 A standing wave produced by the vibrations of a string held fixed at the ends. Notice how the string begins to vibrate from its equilibrium position (*the horizontally dashed line*) until the vibrations reach a maximum displacement (A). The first mode of vibration or fundamental wave is at the bottom of the figure, and the harmonics or multiples are shown above it. The number of nodes (N) and antinodes (A) repeat themselves in sequences as one goes from the fundamental to higher modes of vibration

Conceptual Task

How does the number of nodes and antinodes change as you move up from the fundamental wave to the forth mode or third harmonic?

Which one leads in number as the sequence unfolds?

Exploratory Task

Standing waves can be generated in a variety of ways; we can use water waves to represent a standing wave created by the reflection of the waves from a surface (a wall). Use the online simulation at (http://phet.colorado.edu/index. php); Choose *Wave Interference* from the available choices, select the water tab and make sure the screen looks like the figure below.

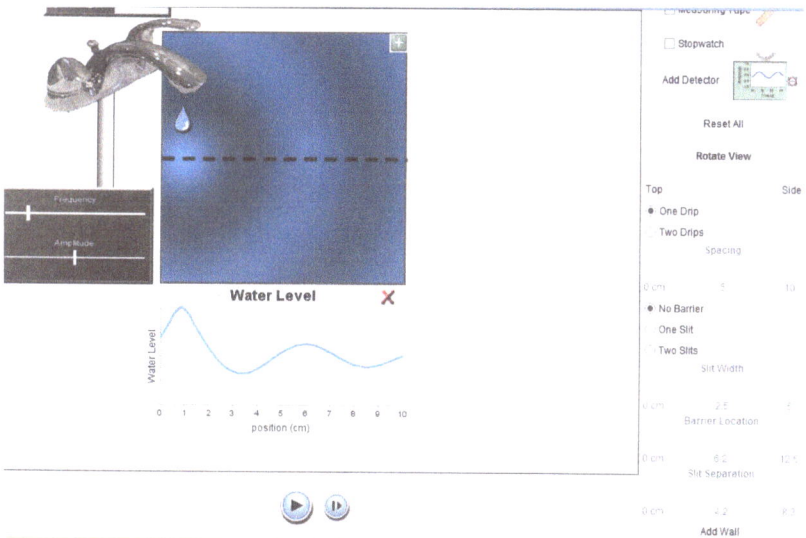

You will notice as you run the simulation that the amplitude of the water wave decreases and we expect that as the waves exit the screen on the right they eventually dissipate.

- Pause the simulation, add a wall, and adjust its position so that it looks like the figure below.

(continued)

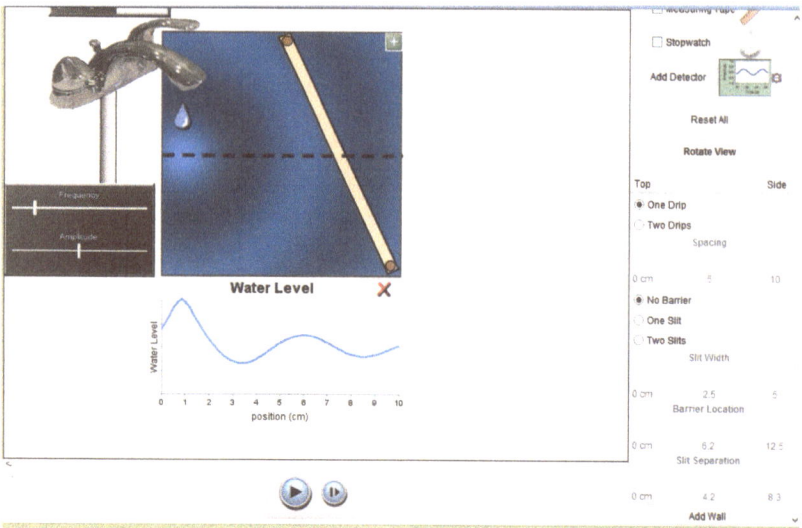

- Continue with the simulation and describe what you see; then gradually increase the frequency until a clear pattern of alternating bright and dark regions appears. Now rotate the wall until it is fully vertical and at the right end of the figure, and wait for a few seconds. How do the representations from the figure and the graph compare?
- How can we conclude that this is an example of wave interference?
- Stop the simulation, click "Reset All" and choose Two drips; make sure the amplitude and frequency are towards the middle of their ranges. Run the simulation and describe what happens; how does what you see now resemble the part with the wall?

The same types of interference can be seen when you choose "Sound" and "Light" in the above task. It is important to notice that the units of intensity in the graphs are different: water level for the water, pressure for the sound, and Electric Field for the light, as these represent different vibrational phenomena that can all display interference patterns.

Beats

When two waves start out with slightly different frequencies and they interact, the result is an alternating pattern of constructive and destructive interference; the pattern exhibits a particular shape called a beat. It has a given frequency, which depends on the initial frequency difference between the interacting waves. Figure 5.2 shows the evolution of the pattern, from its beginning to its completed shape.

Conceptual Task

In the diagrams that follow, the beat frequencies are a function of the initial frequency differences between two interacting waves. In (a) the difference is 50 Hz, in (b) it is 40 Hz, in (c) it is 30 Hz, and in (d) it is 5 Hz.

What is the relationship between frequency difference and beat frequency?

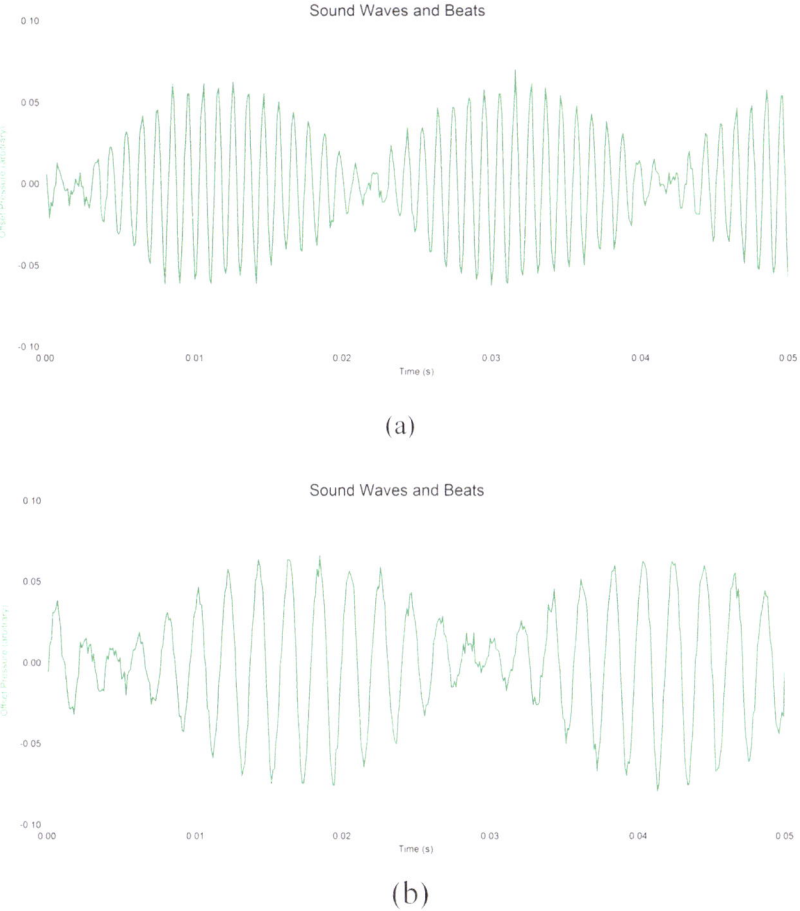

(a)

(b)

(continued)

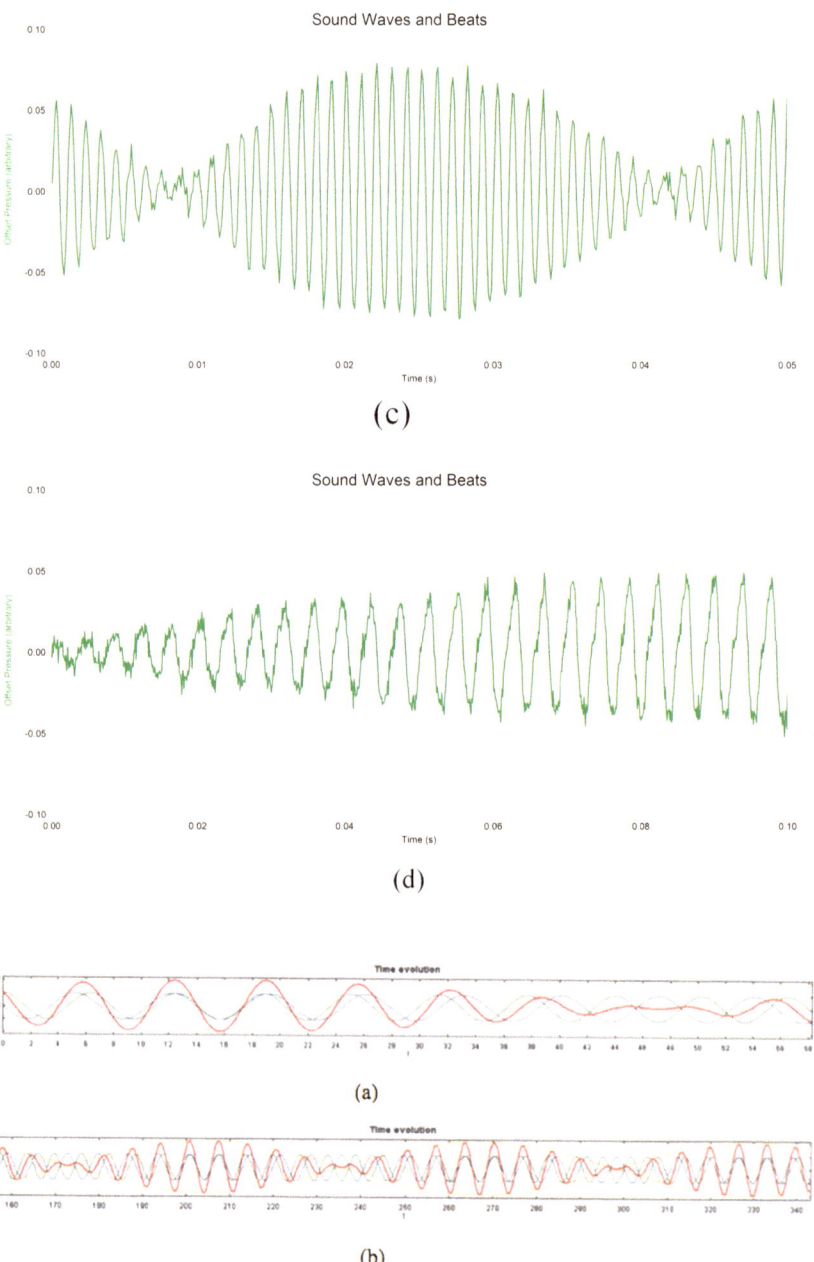

Fig. 5.2 The development of beats occurs as two waves of equal amplitude but slightly different frequencies interact. In (**a**) the interacting waves are shown along with the resulting (larger amplitude) wave; as they begin to change phase, or to be out of step with each other the resulting amplitude decreases. In (**b**) the waves are shown after a sufficiently long time interval as their interaction has developed the pattern of beats. Whenever the waves are out of phase or opposite to one another, the beat will have a node; when they are in phase or in step with one another, the beat has an antinode

Resonance

The phenomenon of beats serves as an introduction to an important property of waves that results when an interaction between two waves results in a maximum value for the resulting amplitude. This is generally known as resonance, and many phenomena as well as systems that display wave properties exhibit it.

Whenever vibrations result in a wave that is acted upon by an external agent such as a variable force that itself has a repetitive pattern, the wave is affected depending on the periodicity of the applied force. Consider a child on a swing set; to get started a push is needed (the external force) and the child moves back and forth in an oscillating pattern that can be described by a wave. As time goes on, the amplitude of the wave decreases (the child will return to the starting position), unless there is a means to keep the oscillations going. This can be provided by continuing to apply the push, or as the child will learn, by tucking and extending its legs. However, regardless of how it is done, it must be done *in unison*, or with the same periodicity as the swinging motion of the child. The resulting successful maintenance of the motion is an example of resonance. The amplitude of the oscillations described by the child's motion is kept at a maximum, thus facilitating the continuation of the swinging activity.

Exploratory Task
Virtual Demonstration of Resonance
   Use the PhEt simulation "resonance" available at:
   https://phet.colorado.edu/en/simulation/legacy/resonance
   Make sure the initial figure looks like this

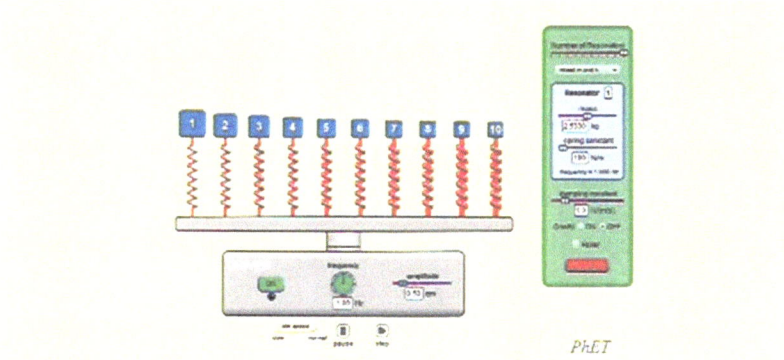

Prediction: Is there a way to make the platform move up and down so that one of the springs oscillates greatly as compared to the others?

1. Choose 2 resonators from the toolbar of the number of resonators
2. Turn the driver on

(continued)

3. Observe what happens for a few seconds; what do you see shown on the simulation that explains why the amplitude of simulator 1 is the greatest?
4. Click on the box showing Resonator 1 and type 2, hit enter and note the frequency of resonator 2 at the bottom of the blue box.
5. Now change the frequency of the driver by rotating the knob under the moving platform and setting it to a value equal to that of resonator 2.
6. What did you observe happening after a few seconds?
7. Finally slide the top rider indicating the number of resonators until you see six (6) springs appear oscillating on top of the moving platform.
8. Again, click on the box showing Resonator 2 and type the value of any of the added oscillating springs. When you see its frequency appear, change that of the rider until it matches it and determine if the same thing that happened before takes place now.
9. Does this observation confirm your prediction?

There are many instances of resonance, and a good activity that exposes us to its devastating effects (quite unlike the successful effect of keeping a child moving on a swing set) is provided in the following task.

Conceptual Task
Consider the situation described by many videos (on you tube) where the Tacoma narrows suspension bridge collapsed. The bridge had been oscillating basically in the fundamental mode of vibration (see the bottom part of Fig. 5.1) for months. However on the morning it fell, a new mode of vibration appeared.
      Can you identify the mode by comparing the parts of the video showing the bridge's motion that day, and one of the modes from Fig. 5.1?

Exploratory Task
Use the PhEt simulation that allows you to display the motion of a mass or a system of masses attached to springs available at:
      https://phet.colorado.edu/en/simulation/normal-modes
      Try to simulate the motion of the Tacoma narrows bridge from the video by using the simulation and pointing out features from the film that can be reproduced with it. Some suggested ways to use the simulation follow:

1. Choose 1 mass, slide the frequency button upwards and then start the oscillation.
2. Choose 2 masses and manipulate the frequency modes until you get the system to move in a similar way to the oscillating bridge.

As seen from the previous tasks there are many interesting applications of the concept of resonance involving a wide variety of phenomena. It is one of those unique wave properties that can provide insights into many other phenomena where such behavior (matching frequencies) may help us see relationships not otherwise apparent. In particular, how information content (in the form of energy or other intensity relations) can be transferred and maximized between systems.

Application to Sound

The quantitative differences between various types of vibrations resulting in standing waves can be understood as these are produced in two ways:

1. By the vibrations of a string held fixed at both ends
2. By the vibrations of an air column in tubes

String vibrations and the vibrations in a column of air in a tube that is open at both ends have similar properties. The following figures show the differences, as well as the common features between these vibrations, and those in a tube with one end closed.

Figure 5.3 shows how the string vibrations and those of air inside a tube open at both ends are similar in the patterns produced. Both show that for the fundamental mode of vibration the wavelength is ½ the length of the string, or the tube. It is easier to see that for the string than for the tube since a half wave is more readily perceived, although both exhibit the property that two corresponding points on the waves shown, separate a linear distance that corresponds to half a wave. As the next three vertical modes of vibration, or harmonics appear, they are simply repeating patterns of the fundamental one. The first four modes are shown, and the difference is that for a string the sequence of the number of nodes moving vertically leads that of the antinodes, whereas it is the reverse for the air vibrations.

Fig. 5.3 The patterns resulting from the vibrations of a string held fixed at both ends, and from air vibrating inside a tube open at both ends. In both cases (N) are nodes or points that are not displaced vertically, and (A) are antinodes or points where the displacement from equilibrium (*the horizontally dashed line*) is a maximum

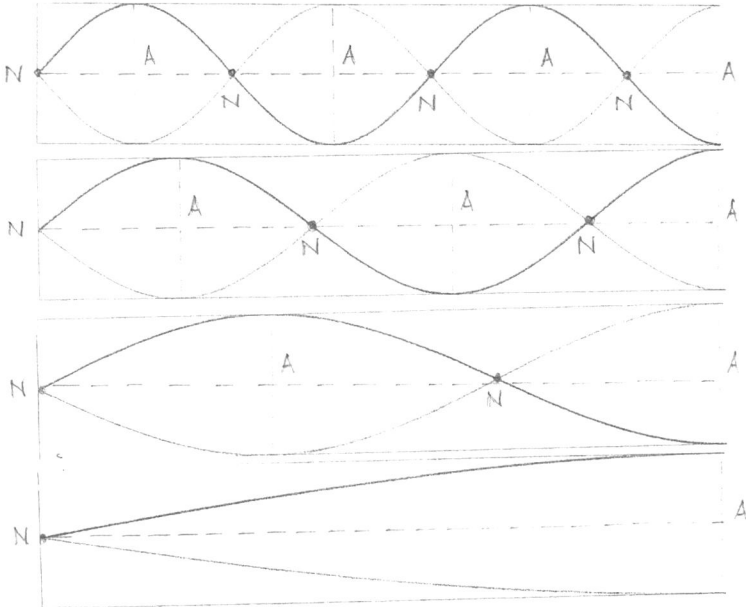

Fig. 5.4 The vibrations are those of air inside a tube with one end closed

The situation is different in Fig. 5.4 where neither the nodes nor the antinodes lead; they are equal in number as the harmonics repeat the pattern created by the fundamental mode of vibration. Additionally, as the pattern is reproduced vertically the need to maintain a node at the closed end, as well as to have a corresponding increase in the number of antinodes requires that only odd multiples appear, there are no even multiples like in Fig. 5.3.

Air vibrations can be understood in terms of either vertical air displacement or changes in air pressure. One can see that they are inversely related, when the air movement is a maximum the air pressure is a minimum. Correspondingly, when the air movement is a minimum the pressure is a maximum.

At this point it is important to remember that the actual motion of air is that of a longitudinal wave, where the maxima and minima are really regions of compression and expansion of air inside the tube. The above representation is only for the purposes of illustrating the similarities and differences between air vibrations and those of a string, which are transverse since the string cannot be compressed, although being under tension generates the standing waves.

The mathematical representation of the above features summarizes the difference between the recurring patterns, and it also illustrates the reason why in the case of sound produced by wind instruments there is a difference of tone quality (timbre) despite the frequency being the same.

For both cases in Fig. 5.3 the fundamental frequency is obtained from $v = \lambda f$ using $L = \lambda/2$

$f_1 = v/2L$, where f_1 is the fundamental frequency corresponding to the lowest mode of vibration.

For the second harmonic or first overtone $L = \lambda$, and so $f_2 = v/L$

Expressing as a ratio $\dfrac{f1}{f2} = \dfrac{v/2L}{v/L} = 1/2$ which means that $f_2 = 2f_1$

For the third harmonic or second overtone $L = 3/2\ \lambda$, and so $f_3 = 3v/2\ L = 3f_1$

Finally, for the fourth harmonic or third overtone $L = 2\lambda$, and so $f_4 = 2v/L = 4f_1$

Consequently, for string and air vibrations in a tube open at both ends, all the harmonics (even and odd) are present.

However, for Fig. 5.4 the fundamental frequency $f_1 = v/\lambda$, and using $L = \lambda/4$ gives

$$f_1 = v/4L$$

For the first overtone $L = 3/4\ \lambda$, and so the next harmonic is $f = 3v/4\ L = 3\ f_1$ or f_3 instead of f_2!

The same is true for the next overtone where $L = 5/4\ \lambda$, and so its frequency is $5f_1$ or f_5

Therefore, we can see that for a tube closed at one end the even harmonics are missing, only the odd ones appear. That is what makes a difference in the tone quality produced by wind instruments that are open at both ends (like a flute), and those reed instruments (like a clarinet) that are open at one end only.

Exercises

1. What is the length of a string that produces a second harmonic frequency of 300 Hz if its wavelength is 20 cm?

   (Hint: this exercise can be solved either by getting the length from the equation for the second harmonic directly or by realizing that the second harmonic is simply twice the first, and then finding the length from the fundamental one).

2. How long must a tube closed at one end be if the second harmonic of air vibrations has a frequency of 600 Hz? (Use $v = 340$ m/s)

3. (A) What is the wavelength for a 2 m long tube open at both ends when the third harmonic has a frequency of 900 Hz? (Use $v = 340$ m/s)

   (B) How many antinodes and nodes exist in this mode of vibration?

Adding harmonics can have interesting applications in sound due to the effects that a set of tones of different frequencies can have upon listeners. One such example is the sound produced by a group of harmonics where one can hear the fundamental frequency mixed in with the higher overtones; remarkably, a collective sound produced by such a group will contain a tone perceived by the listener as the fundamental, even if that frequency is missing from the group!

Exploratory Task
Use the PhEt simulation "Fourier: Making Waves" available at
  http://phet.colorado.edu/en/simulation/fourier
  Choose the first eight harmonics so that each of their amplitudes is 0.5;
then click the speaker symbol to hear the sound. After a few seconds, turn off
the fundamental (A_1) by lowering its amplitude to zero, and then listen carefully to detect its presence amid all the remaining frequencies heard.

Application to Light

The combination of waves with varying frequencies and wavelengths can also occur
in the case of light, where the composition of the spectrum can help understand how
colors are formed and perceived. Red, green, and blue are the primary colors of
light, which correspond to color-sensitive cells called cones in the human eye; yellow, magenta (a purple color), and cyan (a green-blue color) are called secondary
colors. Each color is associated with a particular wavelength.
  The classic demonstration of the decomposition of white light into its respective
colors is attributed to Newton, who also demonstrated that not only can a beam of
white light be split into various colors by a glass prism, but another prism can be
used to recombine them into the white light initially observed. Figure 5.5 illustrates
this phenomenon.
  Color mixing follows a similar type of property to that of sound frequencies,
where a net result is obtained from adding or subtracting individual frequencies to
produce specific patterns of sound. In the case of light the addition of the fundamental or primary colors (red, green, and blue) will in turn produce secondary colors
that are a mixture of those. This is the way television and computer monitor colors
are produced, where the middle area (white) corresponds to a pixel. For subtractive
color mixing one begins with the secondary colors (cyan, magenta, and yellow) and
by successive filtering arrives at the absence of color (black). This is the way print-
ing and painting color mixing is done (Fig. 5.6).

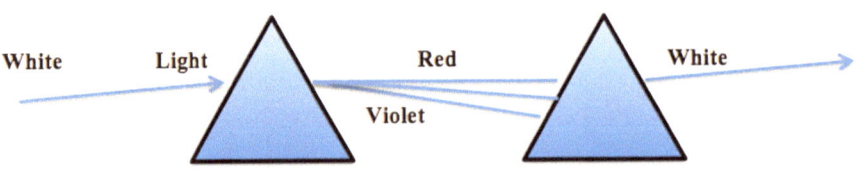

Fig. 5.5 Two prisms are used; one to separate the white light into its colors, and another one to
recombine the spectrum into the original white light

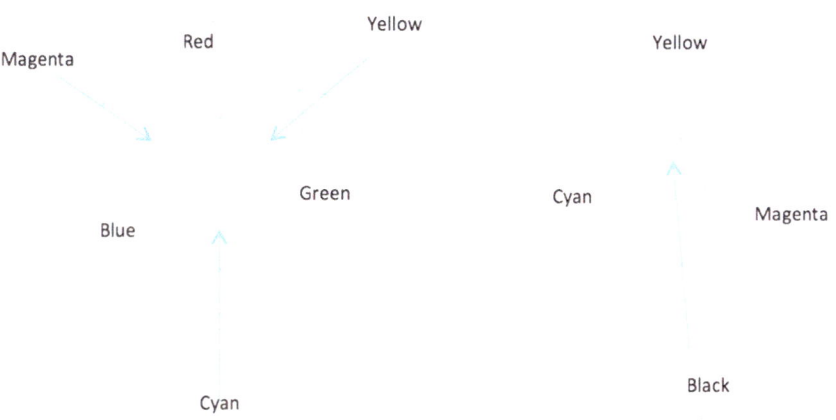

Fig. 5.6 Color mixing of the primary and secondary colors by addition and subtraction. The center of the figure on the left would be white color

Exploratory Task

The two processes of color mixing can be demonstrated with the use of the PhEt simulation

"Color Vision" available at http://phet.colorado.edu/en/simulation/color-vision

1. For additive color mixing select RGB bulbs and gradually change the intensity of each of the colors; observe the secondary colors that are perceived as the intensity of the primary ones changes. In particular, notice the shades produced as any given two primary color intensities are gradually increased and mixed.
2. For subtractive color mixing, first select the white light and notice how the color perceived corresponds to that of the filter as you slide the filter control. This corresponds to the way colors are perceived when white light shines on an object that has been painted a given color, where the other colors of the white light are absorbed, and the one corresponding to the painted one is reflected.
3. *Demonstration of Resonance*— Now select the colored bulb and notice that as you change either the bulb or the filter color, the perceived color is always black unless the two colors match. This is the result of the two frequencies or wavelengths being identical, which produces the maximum intensity of the light perceived.

Fig. 5.7 A standing wave
created by a vibrating
string, containing more
than one wavelength

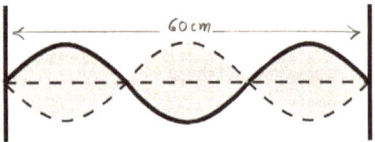

Problems

(1) (a) How long should a *closed* tube in air be such that its fundamental
         frequency is 200 Hz?
         Use $v = 345$ m/s.

    (b) What is the frequency of its next possible mode?

(2) The second possible standing wave in a *closed* tube in air has a frequency
    of 1000 Hz.

    (a) What is its fundamental frequency?
        Use $v = 345$ m/s.

    (b) What is the fundamental wavelength?

    (c) What is the length of the tube?

(3) (a) A standing wave is oscillating at 690 Hz on a string, as shown in Fig. 5.7.
         What is the wavelength?

    (b) What is the speed of traveling waves on this string?

    (c) How many nodes and antinodes are shown in Fig. 5.7?

Chapter 6
Diffraction

We already discussed one type of bending experienced by waves as they are refracted upon encountering media with different indices of refraction. In that case we saw that when a wave goes from a medium of lower to one of higher index of refraction, the rays describing the wave bend toward the normal; when the rays emerge from a medium of higher to one of lower index of refraction, they will bend away from the normal. The property called diffraction is another bending of sorts, but it occurs as waves travel through a medium, and encounter an obstacle. Obstacles can vary in shape and size, and the bending is always dictated by their geometry.

The first part of this chapter is concerned with light since there are many interesting features of the historical development of the understanding of diffraction while using light, as well as many interesting applications. We shall then apply the concepts and principles to sound.

A good example of the ability of light and sound to diffract around obstacles, but for one of them not to be readily perceived, is provided by the following situation. Imagine that you are communicating with someone, and that communication consists of both audio and visual information. If you were walking backwards (very carefully!) while having the conversation, and then you go around an obstacle such as a large tree, you will continue to hear the person but not see him/her while you are behind the tree. In fact, the situation can be made even more dramatic by having you turn around a corner of a building and still notice the same thing. The reason why you can hear the person but not see them is that the size of the obstacle is more comparable to the wavelength of sound, than to that of light; while both light and sound are diffracted by the obstacles the effect isn't noticeable for light (you do not see the person), but it is for sound (you can hear them).

Conceptual Task
Suppose in the situation just described, you were also in communication with the other person by using your cell phone; as we know the signals are electro-magnetic (like light), and yet you can still receive them after you have lost the visual information. Why you do suppose this happens?

© Springer International Publishing Switzerland 2017
F. Espinoza, *Wave Motion as Inquiry*, DOI 10.1007/978-3-319-45758-1_6

Exploratory Task

Use the online simulation at (http://phet.colorado.edu/index.php); Choose *Wave Interference* from the available choices, then select the Water tab, make sure the screen looks just like the figure below:

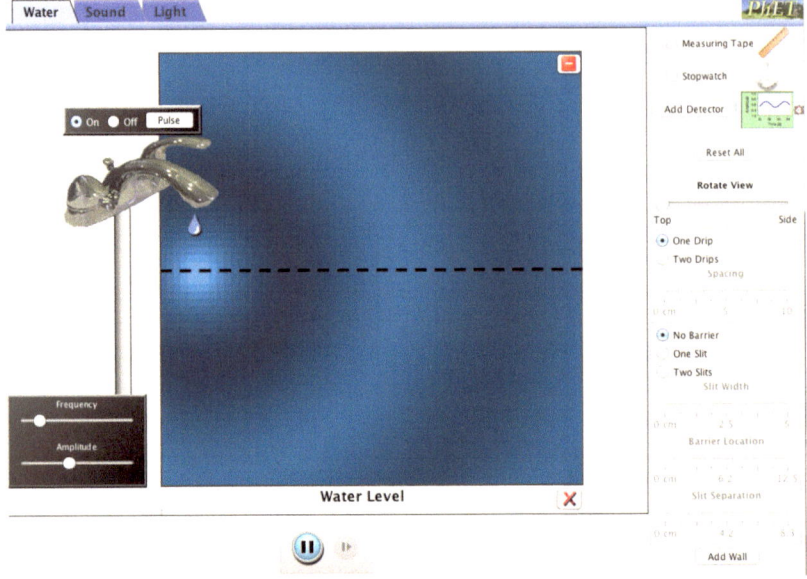

Describe what you observe as the droplets fall

Now click on the Add Wall button and orient the wall so that is appears vertically as shown below

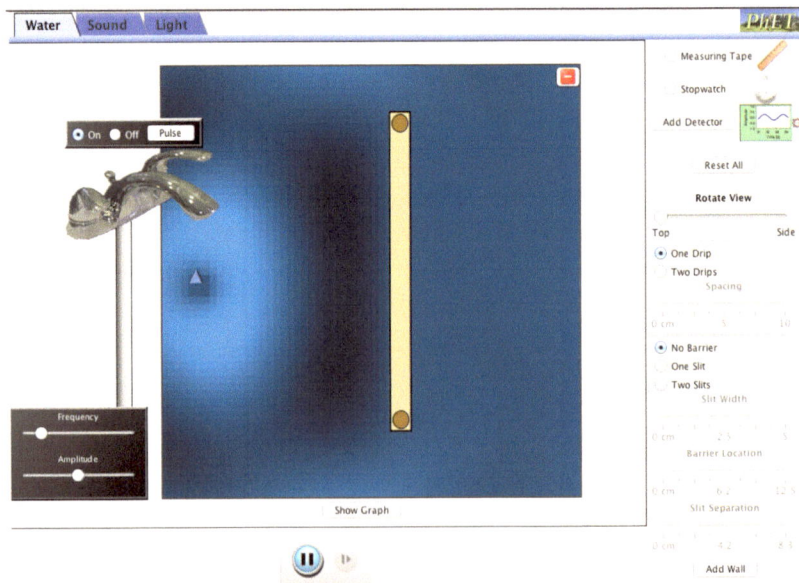

(continued)

Describe what you observe as you slide the wall across the screen, and as you shorten its length.

The important aspect of diffraction is that the wavelength of the diffracting waves must be comparable in size to the dimensions of the obstacle, as determined by its discoverer Francesco Maria Grimaldi in the seventeenth century. He observed that when a beam of light came into a dark room and projected onto a surface, the spot made resembled the opening through which the light came in. If the opening is circular, the spot is also circular; however, as the size of the opening decreases the spot at some point will be surrounded by a set of concentric circles. This is the result of the wave bending around the opening and reproducing its shape onto the projection.

Figure 6.1a shows a set of wave fronts represented by the parallel lines, that upon encountering the gap will diffract. Notice that in part (b) there are lines drawn past each gap, but these don't imply that there will be projections onto the screen where dots will appear. What will appear as a central maximum representing constructive interference will be directly across from the middle of the line that separates the two gaps, where the diffracted waves are intersecting each other.

The essential aspects of diffraction can be demonstrated by a diagram that illustrates wave fronts being diffracted by a circular opening, and the resulting interference being projected onto a screen at a distance from the opening. This is shown in Fig. 6.2.

Figures 6.2 and 6.3 show the details of diffraction as wave fronts incident from the left go through a circular opening. The size of the opening between points 1 and 2 is taken as the diameter and two points on a given wave front are shown illustrating that there are two paths taken between the opening(s) and the screen where the

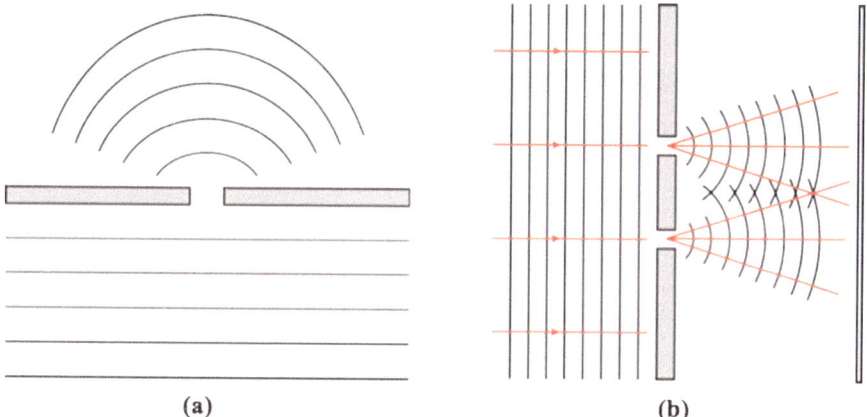

(a)   (b)

Fig. 6.1 The phenomenon of diffraction can be shown as occurring whenever waves encounter obstacles, such as the gap in (**a**) the transmitted wave now diffracts as shown by the bending. As shown in (**b**) two such gaps will lead to interference between the diffracted waves, resulting in alternating constructive and destructive regions of interference

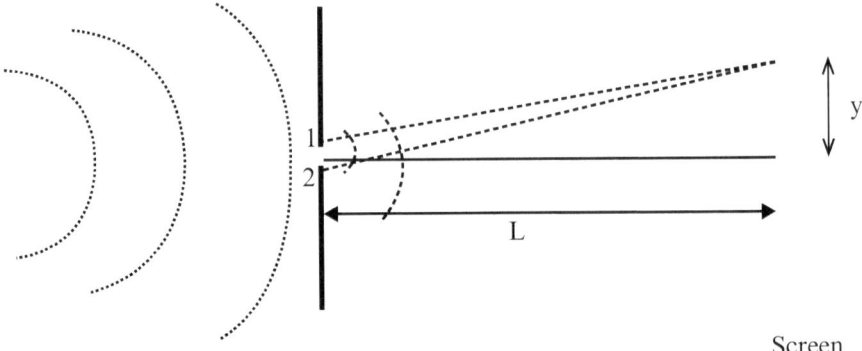

Fig. 6.2 A series of wave fronts on the left are incident upon a circular opening between points 1 and 2, and are shown diffracted as they emerge from the opening. The diffraction causes points on the wave fronts to interfere, forming patterns of constructive and destructive interference on a screen at a distance L from the opening. The distance y is used to illustrate points of constructive or destructive interference

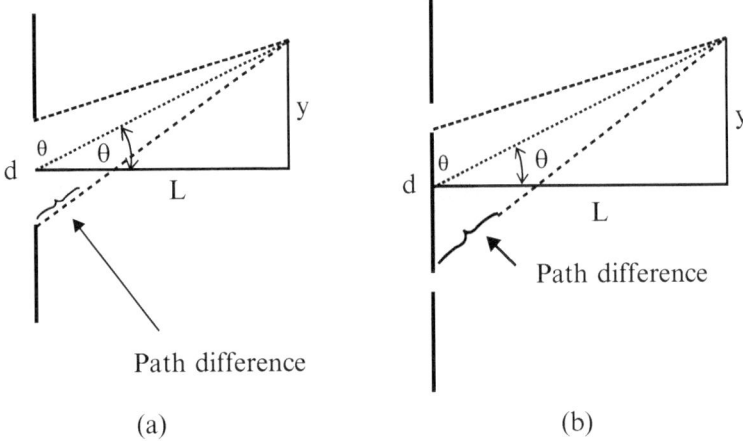

Fig. 6.3 Part (**a**) shows a magnified view between points 1 and 2 of the path difference between two points on a wave front for a single circular opening of diameter d, and part (**b**) shows it for points at the middle of two such openings separated by a distance d

interference pattern can be displayed. The two paths are clearly of unequal length; using a right triangle one can see that the sine of the angle θ is the opposite side (the path difference) divided by the adjacent side (d). Therefore, the path difference is given by $d \sin \theta$.

Now different conditions can be imposed on this path difference to illustrate the criteria for constructive and destructive interference. Notice that there is no path difference along the horizontal line that denotes the distance between the opening(s) and the screen L; this means that for both a single opening and multiple ones, there will always be a bright spot or region along the horizontal representing constructive interference, called the central maximum.

What is typically done with a single opening is to impose a condition for minimum intensity resulting from destructive interference on both sides of the central maximum; this will occur whenever the path difference is equal to one half wavelength. This means that the two paths are out of phase when they arrive at the screen, and thus cancel each other out. For this to happen one can construct a similar triangle by taking half the diameter of the opening to form the hypotenuse, so the path length will be

$$d/2\sin\theta = \pm m\frac{\lambda}{2}, \quad \text{where } m \text{ is } \pm 1,2,3,\dots \qquad (6.1)$$

(The negative values of the integers are only meant to illustrate the fact that the positions of the dark fringes representing destructive interference on the screen appear above and below the central maximum along the length L).

The absence of $m=0$ indicates that there is no central minimum; instead there is a central maximum. The expression then can be simplified to

$$\sin\theta = \pm m\lambda/d \qquad (6.2)$$

Equation (6.2) states that for a single opening diffraction, there will be a series of regions of destructive interference on both sides of the central maximum.

For the case of diffraction by two openings the equation for the path difference can be written using Fig. 6.3b as

$d\sin\theta = m\lambda$ for constructive interference, and $d\sin\theta = m\lambda/2$ for destructive interference. Constructive interference of course results when the path difference is equal to a complete wavelength, which means the two interacting waves are in phase.

From Fig. 6.3 one can get

$$\tan\theta = \frac{y}{L} \qquad (6.3)$$

A useful technique called the *small angle approximation* is based on the following relationship

$\tan\theta = \dfrac{\sin\theta}{\cos\theta}$ and since for small values of θ we have $\cos\theta \approx 1$, whenever θ is small

$$\tan\theta \approx \sin\theta$$

And so $d\sin\theta = \dfrac{y}{L} \qquad (6.4)$

Equation (6.4) can be used to determine various parameters, especially when a multitude of openings are used in a device called a diffraction grating. Such a device is used to split white light into its colors, and has a great number of important applications. The openings do not necessarily have to be circular, in the case of diffraction gratings they are slits resulting from the stretching of a plastic material, and many of these slits are packed into an extremely short width of material where the value of d is typically expressed in microns (millionths of a meter in length).

Example

Consider the situation where laser light goes through an opening that is 0.25 mm wide, and a central maximum appears on a screen that is 100 cm away. If the width of the central maximum is 4.0 mm, what is the wavelength of the light?

We use the small angle approximation, and begin with $m=1$ for the central maximum.

Then using $\sin \theta = \pm\, m\, \lambda/d \rightarrow \sin \theta = \lambda/d$

And $\tan \theta = \dfrac{y}{L}$

Since the width is 4.0 mm (this represents $2y$ from Figure 3a), then $y=2.0$ mm, and $L=100$ cm

Converting to meters: $y=2.0\times10^{-3}$ m, $L=1.0$ m, and $d=2.5\times10^{-4}$ m

$\tan \theta = 2.0\times10^{-3}$ m$/1.0$ $m=0.002$

And since $\tan \theta \approx \sin \theta$

$\sin \theta = \lambda/d$ becomes $\lambda = d \sin \theta = (2.5\times10^{-4}$ m$)\,(0.002)=5.0\times10^{-7}$ m

Such short wavelengths are usually expressed in nanometers (nm), and 1 nm $=1.0\times10^{-9}$ m

Hence $\lambda = 500$ nm.

Conceptual Development Task

As stated earlier the phenomenon of diffraction depends on the scale of the dimensions of the obstacles that cause waves to exhibit such a property. One can explore the way in which the size of obstacles such as slits can affect the way light diffracts. The setup is shown in the figure below:

(continued)

1. Use a source of white light with the glass slab containing a variety of slits and notice the way the pattern appears on the screen as you let the light through each slit. What do you notice as the number of slits increases?
2. Repeat the procedure but instead of white light, use a laser of a given color or wavelength; make sure not to look at the reflection of the laser on the glass plate, and instead concentrate on the patterns produced on the screen for each type of slit. What is different now from what you observed in part 1?

When white light passes through a diffraction grating, the spectrum of colors can be seen. When monochromatic light (light of a single color, such as that produced by a laser) passes through a diffraction grating a series of dots of the same color will instead appear on a screen. These are the regions of constructive interference, and the space between them corresponds to regions of destructive interference.

The following figures illustrate the cases of diffraction due to a single opening, as well as due to two such openings.

Figure 6.4 shows the diffraction of light caused by a single opening. A source sends the wave fronts from the left and when these encounter the obstacle (the single

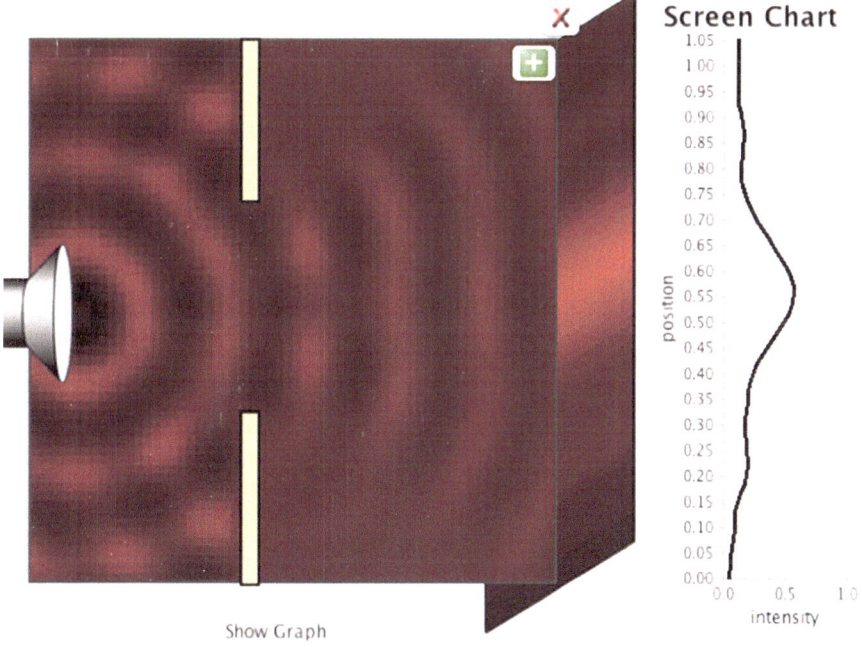

Screen

Fig. 6.4 Diffraction caused by a single opening. The source of light sends circular wave fronts that are diffracted by the opening and interfere to produce the pattern shown at the screen as a central maximum. Note that the intensity graph shows the largest amplitude (at 0.55) for the central maximum, and two smaller peaks (at 0.87 and 0.22, respectively)

opening) they are diffracted. The left side of the figure shows the wave fronts containing regions of light and dark bands, which is due to the wave fronts interfering with themselves upon reflection from the obstacle. One can see how the wave fronts bend around the edges of the opening and propagate toward the screen, where the pattern can be displayed.

The peak of intensity corresponds to the region of maximum brightness on the screen, and the gradual change in the curve corresponds to the decrease in brightness, as you go up and down from the central maximum.

Exploratory Task
To explore the use of the equations expressing the relationships between the various factors involved in diffraction

$$\sin\theta = \lambda / d \text{ and } \tan\theta = \frac{y}{L}$$

And using the small angle approximation $\sin\theta \approx \tan\theta$, one can write the following equation:

$$\frac{\lambda}{d} = \frac{y}{L}$$

To determine how the width of the central maximum changes, one solves for y

$y = L \dfrac{\lambda}{d}$, and so the width $(2y) = 2L \dfrac{\lambda}{d}$

Use the online simulation at (http://phet.colorado.edu/index.php): Choose **Wave Interference** from the available choices, then select the Light tab, make sure the screen looks just like the figure below:

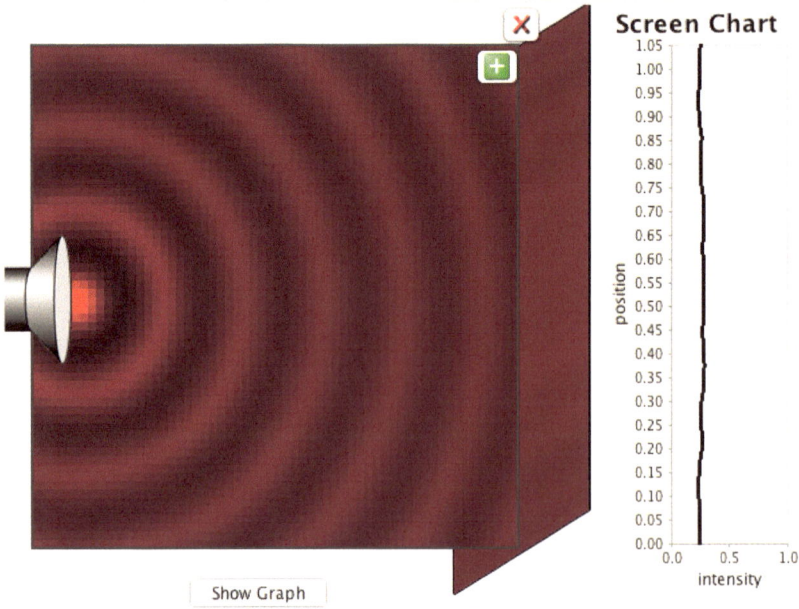

(continued)

The objective is to test the dependence of the width of the central maximum, on each of the variables in the equation (L, d, and λ). Change one variable at a time, and describe what happens as you reach the limits of the values of each variable in testing the relationship

Width of central maximum $= 2L\,\dfrac{\lambda}{d}$ (the 2 is a constant factor that does not affect the relationships).

The use of a single opening or slit for diffraction applications is important as it tells us what are the limits of imaging (used in a general sense, not just with light). The dependence of the width of the central maximum on the various factors explored in the previous task can be used to show why one cannot image an object that is smaller than the wavelength of the signals used.

In the equation "Width of central maximum $= 2L\,\dfrac{\lambda}{d}$" if one uses a device where L and d are constant, the only variable left is the wavelength λ. The usual condition for single slit diffraction is that $L \gg d$ (the distance from the slit to the screen is much greater than the width of the slit).

Recall that the range of visible wavelengths is approximately (400–750 nm); if we were to use light of 600 nm being incident on a slit ¼ mm wide as used in a previous example, $d = 2.5 \times 10^{-4}$ m, and projected onto a screen at 2.0 m distance, we would get a central maximum of width $2y$ where

$2y = 2\,(2.0\text{ m})\,(6.0 \times 10^{-7}\text{ m})/2.5 \times 10^{-4}\text{ m} = 0.0096$ m or 9.6 mm, which is significantly larger than the width d.

If we were to use instead a width an order of magnitude greater (ten times), $d = 2.5$ mm
$2y = 2\,(2.0\text{ m})\,(6.0 \times 10^{-7}\text{ m})/2.5 \times 10^{-3}\text{ m} = 0.00096$ m or 0.96 mm.

This represents a central maximum width that would be smaller than the width d of the slit, and being nearly 1 mm it would be extremely challenging to see. We can compare this result with the first part, where the central maximum width is almost 1 cm and thus definitely visible.

However, there is an additional problem with the second result besides the fact that it becomes very difficult to see the central maximum width. The following figure is designed to illustrate the results of the above exercise, not to scale of course.

As Fig. 6.5 shows, in the first part (a) there is a definitely clear central maximum with a peak that gradually goes to zero, as expected for the adjacent minima. In part (b) the central maximum is no longer there, instead there is a tendency for a central minimum to form, although the intensity does not go to zero before repeating the pattern. The second pattern almost seems like a reversal of the first part; instead of being sharp at the middle, the pattern is now blurred. In addition, the distance between minima has increased so much that there is no way to tell what it is in part (b) as the pattern essentially goes off scale, and in part (a) it was easily determined to be 0.50.

The dramatic change in pattern shown above results from the change in size of the slit width, even though only one source of light is sending the waves that are diffracted. Looking at the two peaks one may be tempted to extrapolate back and

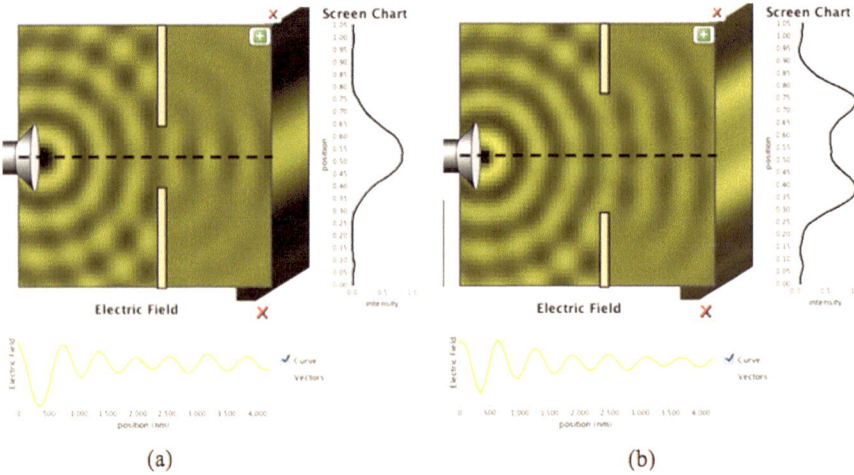

Fig. 6.5 Graphical representations of the changes made in the exercise above. The wave is represented by the changes in the electric field that is part of the electromagnetic oscillations that are light. The figure shows in both parts that the wavelength λ of the light, and the separation L between the slit and the screen remain constant. In part (**a**) there is a clear central maximum with a sharp peak and gradual decrease to zero intensity for the adjacent minima. In part (**b**) the slit width is made significantly larger than in (**a**), and this leads to the fuzzy pattern that now forms where the central maximum was, and the intensity of the peaks does not decrease to zero in between them

imagine that instead of one light source, there are now two. If this effect arises from only one source of light, imagine what happens where there is more than one.

The above result can be used to explain a limitation that exists for optical devices due to the wave nature of light. Imagine two sources of light quite distant from a slit of width d. If light were not diffracted and interfere with itself as it bends around the opening or slit, we would expect to see two bright spots on a screen corresponding to each of the light sources. However, as a result of interference each light source will appear as a central bright spot (the central maximum width) surrounded by adjacent dark areas, and even other maxima further away from the central one. With two light sources being diffracted the result is now an addition of two patterns appearing on the screen. If the two light sources are sufficiently separated their patterns can appear as the result of Fig. 6.6a. However, what one usually gets is the result of Fig. 6.5b, where the two overlap.

Figure 6.6 illustrates what happens in the extreme case where two sources of light cannot be resolved by an optical instrument such as a telescope, or the human eye due to the fact that the interference patterns produced by each source combine. The two maxima overlap and what should happen in (a) instead often results in (b).

Ideally the central maximum of one image should appear where the first minimum of the other image is, a condition or limit of resolution called *Rayleigh's criterion*. This applies to optical instruments in terms of the limit of resolution of images of distant objects, such as stars. However, ordinary objects like automobile headlights can also appear as a single source when they are far away, and only become resolved as two as the distance decreases sufficiently for the human eye to see them as two.

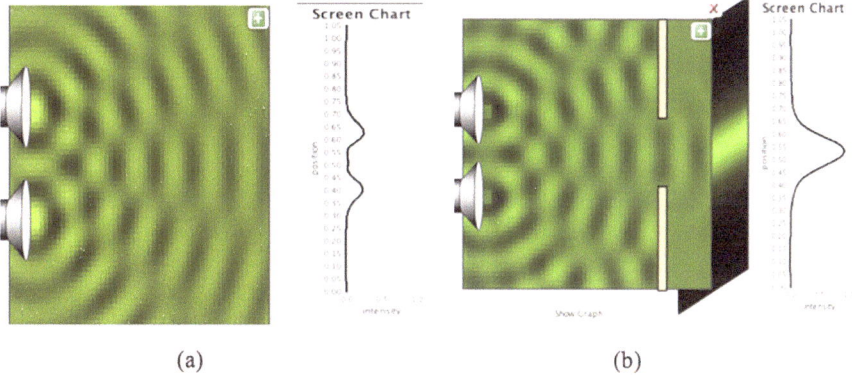

(a)   (b)

Fig. 6.6 Two sources of light sufficiently separated by a distance should exhibit the pattern shown in (**a**) that corresponds to the projected images being resolved, since the two patterns are clearly distinguished. What one usually observes is the result of Fig. 5b where the two patterns overlap. The extreme case is shown in (**b**) where the projected pattern or image appears as that of one, when in reality there are two sources

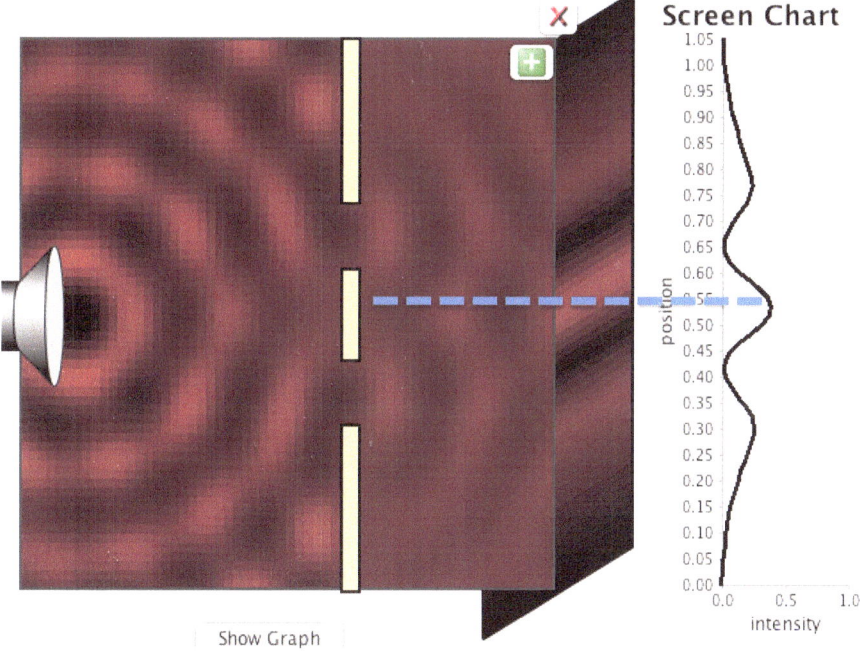

Fig. 6.7 Diffraction caused by two openings. The interference pattern is more discernible than for a single one

Figure 6.7 shows the situation for two openings, with similar features to those of a single one. However, there is more detail available in the pattern. The intensity peak at approximately 0.53 represents the central maximum, those at 0.43 and 0.65 in turn represent the first two minima, and those at 0.30 and 0.77 represent the first-order maxima.

1. The intensity graph representation corresponds to the bright and dark fringes seen on the screen, with the central maximum illustrated by the dashed line. The fact that the central bright maximum appears exactly behind the line separating the two openings serves to illustrate the way the wave model of the nature of light was demonstrated in the early nineteenth century by Thomas Young [1]. If light consisted of particles, as Newton and others had thought, the interference pattern would not be formed since the particles could only appear on the screen directly across from the openings. In other words, there would only be two bright spots on the screen, exactly where the particles impacted it.

Experimental Task

As we saw in the previous activity that used a laser, we want to minimize the exposure to the reflection of the laser light from a glass or other smooth surface. In that task we concentrated instead on the pattern of diffraction on a screen produced by the light as it went through the slits. However, this same property of reflection can be used to determine the wavelength of a given light color. As you may have previously noticed, and as can be seen in the figure below the reflection of light from a compact disc can produce a color spectrum since its surface appears smooth, but it really isn't. There are a number of versions of this experiment available online as well as other excellent sources such as the one described in The Physics Teacher [2].

The objective of this experiment is to use a known wavelength from a monochromatic source (a laser) to determine the diffraction properties of a compact disc, so that when a light source that contains all the colors shines on the disc, individual color wavelengths can be isolated and measured.

Procedure

(I) Determination of the track separation on the CD

In this part we use a laser and shine it through a hole on a screen (a sheet of hard paper) so that it reaches the compact disc (CD) along the direction of the normal to the disc, and it reflects back onto the screen. The screen should be located approximately 20 cm from the CD (***Important: Make sure the laser hits part of the CD tracks, not the middle of the disc***).

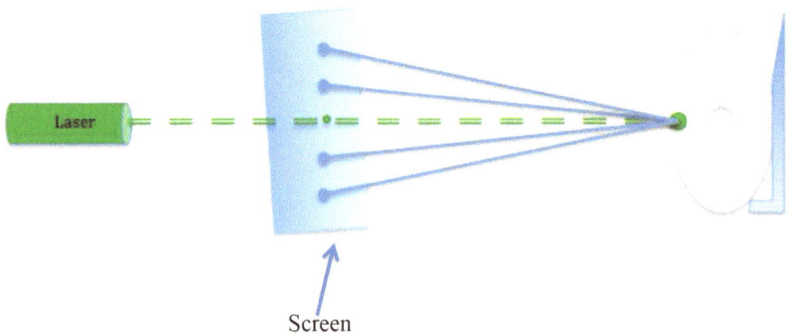

Screen

(continued)

Experimental setup showing the laser going through the slit on the screen and its reflection on the CD forming the set of dots on both sides of the central maximum. The following pictures show the pattern obtained.

You should observe first- and second-order spots on both sides of the center hole; measure the distance D between the inner two (first-order) spots, as well as between the second-order spots.

The diffraction pattern is shown in the dark on the first picture, and with some light on the second, where the laser can be seen to the left of the screen. There are clearly two sets of bright spots on either side of the central maximum (the middle spot); these correspond to the first- and second-order maxima, respectively. Notice particularly in the dark picture that one of the bright spots is slightly beyond the edge of the screen, and the spot that is not part of the pattern on the screen is where the laser strikes the CD.

The geometry of the situation is shown in the following diagram that describes the diffraction pattern in detail.

(continued)

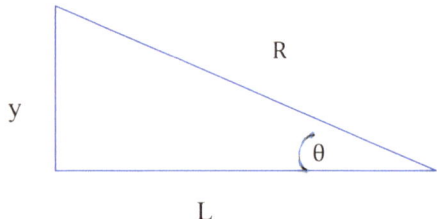

where y is the distance between the central and the first-order maximum; L is the distance between the CD and the screen, θ is the angle formed by the reflected rays for the same two maxima, and R is the hypotenuse of the right triangle formed by these distances.

$\text{Sin } \theta = \dfrac{y}{R}$, and using the Pythagorean theorem, $R^2 = y^2 + L^2$

So $R = \sqrt{y^2 + L^2}$, Hence $\sin \theta = y / \sqrt{y^2 + L^2}$

And using Eq. 6.2

$\sin \theta = \pm m \lambda / d$, we solve for d (which in this case corresponds to the spacing between the CD tracks)

$d = \pm m\lambda / \sin\theta$

since we are using the first-order maximum $(m = 1)$

$d = \lambda\sqrt{y^2 + L^2} / y = \lambda\sqrt{1 + L^2 / y^2} \ (\text{A})$

the two distances y for the first-order maximum should be averaged before using the value of y in the equation.

(II) Determining the wavelengths of chosen colors from the visible spectrum

In this part a light source is placed behind and above an observer who then moves the CD gradually from a distance of roughly 10 cm in front of the face, making sure that a spectrum appears on the surface of the disc. You should practice moving the disc slowly back and forth to confirm that the spectrum appears more than once, since contrary to the laser, instead of dots representing the maxima the light source now produces spectra that appear at values of $m = \pm (1, 2, 3, \ldots)$ from the center of the disc. It is imperative that the disc surface be kept perpendicular to the distance from the eye so as to ensure consistency in the way the spectra appear. When a spectrum is clearly obtained, move the disc until the particular color region chosen as the wavelength to be determined, is just at the outer edge of the disc.

Additionally, in this part the distance from the light source to the observer should be about 10–20 times that of the distance between the observer and the position of the CD.

(continued)

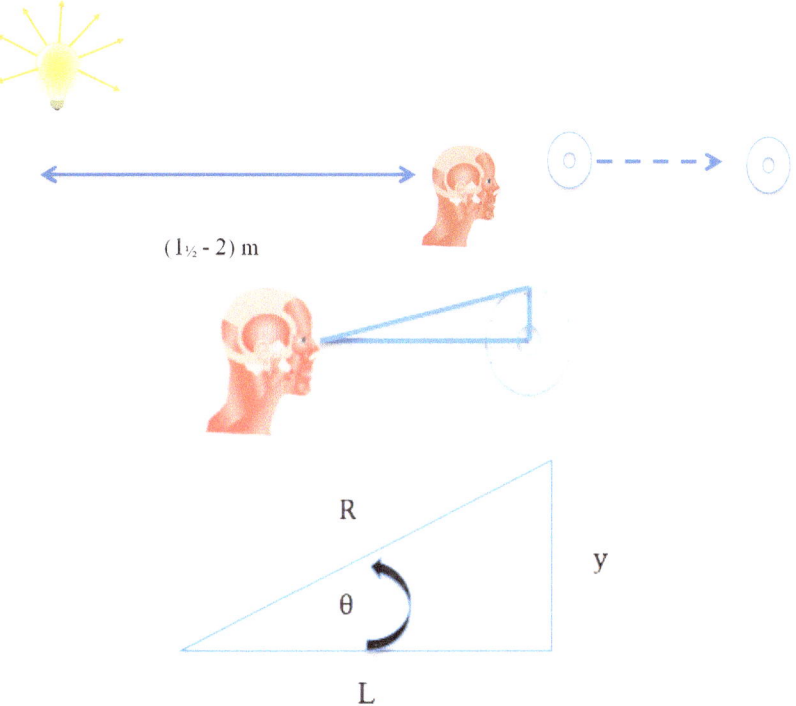

Experimental setup to obtain the desired spectrum from the light on the CD.

When the disc is at the point where the chosen color (wavelength) is on the outer edge of the CD, the geometry in detail is shown below

The relationships used before are still valid, except that instead of measuring y directly we must use instead the diameter of the CD, since the central maximum falls on the hole of the disc. Thus

$\sin\theta = \dfrac{y}{R}$, and $y = D/2$ (half the CD diameter)

And using the relationships from before

$\sin\theta = D/2 / \sqrt{D^2/4 + L^2} = 1/\sqrt{1 + 4L^2/D^2}$

and $\sin\theta = \pm\, m\,\lambda/d$, with $m = 1$

(continued)

sin $\theta=\lambda/d$, using the value of d obtained in (A) of Part I we can now find the wavelength

$$\lambda = d\sin\theta = d / \sqrt{1+4L^2 / D^2}$$

Write your calculations here and then fill in the value of the wavelength in the table below

Color	Wavelength	Range (nm)
Red		620–780
Violet		390–455

Reflections

(1) What values of the ranges are your wavelengths closest to?
(2) What do you consider the main sources of error in this experiment?

Application to Sound

The bending of waves around obstacles also takes place when sound propagates, and as with light there are many applications. For example, if you are listening to sounds coming from a room through an open door, you will have noticed that you don't have to be standing in the middle of the opening to hear the sounds. Even if you walk away from the door, you will still be able to hear the sounds due to the bending of sound waves around the obstacle represented by the door opening. Additionally, one can hear the lower frequencies of sounds coming from ensembles such as a marching band, before one hears the higher pitch sounds. This is a result of lower frequencies bending or diffracting more than higher ones.

But why is this so? Generally speaking waves will diffract whenever their wavelengths are longer than the size of the obstacles that cause them to diffract. That is why the initial example in this chapter results in one being able to hear around obstacles, but unable to see around them. The wavelengths of light are much smaller than the sizes of the objects causing the diffractions in the example, so they end up being reflected rather than diffracted by objects ordinarily encountered everyday.

Consider the example of a loudspeaker producing sounds of various frequencies within the audible spectrum (20–20,000 Hz); choosing two from among the range, such as 200 and 2000 Hz yields the following values for their wavelengths:

If we assume the speed of sound to be roughly 340 m/s, the 200 Hz sound will have a wavelength given by

$$v=\lambda f, \text{ and so } \lambda=v/f= \frac{340\text{m}/\text{s}}{200\text{ Hz}} =1.7\text{ m}$$

whereas the 2000 Hz sound will have a wavelength

$$\lambda = v/f = \frac{340 \text{ m / s}}{2000 \text{ Hz}} = 0.17 \text{ m or 17 cm}$$

A typical loudspeaker capable of producing such sounds will definitely be larger than 17 cm and so will send out waves that will spread out more for the longer wavelengths than for the shorter ones. This particular property will be explored in the next experimental task.

Exploratory Task

Contours are graphical representations of data where the values of certain quantities are constant, but there are also changes between the contours themselves. You may have noticed contours appearing on weather reports showing the values of temperature or air pressure. There are many possible uses of data collected and displayed as contours.

One important property of contours is that they are constructed from data points displayed and connected whenever they have the same value. Since contours represent numerical data they can overlap, but not intersect. In other words, unlike waves contours cannot go through each other. The following figure illustrates what contours look like in general.

The two contours shown indicate values that are the same for each curve, but the curves themselves differ in value. The changes in value could be an increase or a decrease, and can be shown changing inward or outward in the figure.

In this experimental task measurement and representation as contours (lines of constant value) for sound intensity (in dB) as two tones of different frequency are emitted, and as the measurements are taken progressively farther from the source.

Using a sound level meter (either a microphone or a phone app) one can determine on a predetermined grid what the pattern of propagation for sounds of different frequencies is, to determine which one diffracts more.

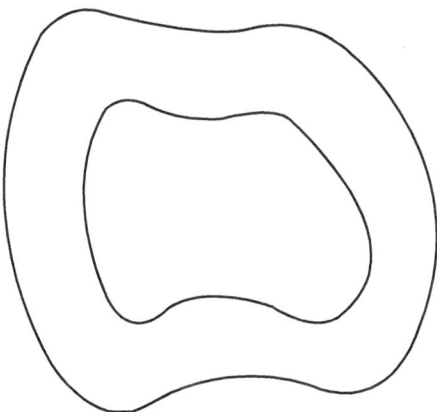

(continued)

1. Sound of a given frequency is produced; determine a location where the loudness level of sound has a specific value in dB, and repeat the measurement for other locations that have the same loudness level.
2. Mark the locations on a grid and then translate the physical measurements onto a model on paper where the points having the same loudness level can be connected with a line, this will be the first contour.
3. Repeat the measurements for sound of a higher frequency than the first one.
4. Construct the second contour and compare the spreading with that of the first one. What do you notice that is different in the spreading?

Inquiry-Based Investigation

We are familiar with many examples of the use of sound in nature by other species besides humans. From insects to whales there are many accounts and descriptions of their use of some of the properties exhibited by sound waves, particularly for navigation, hunting, and echolocation, to name a few. It is known that elephants and bats, to choose two examples from the animal kingdom, use ultrasonic (beyond the higher limit of the audible range), and infrasonic (beyond the lower limit) for communicating and/or hunting purposes.

1. Determine the range of the dimensions of typical objects animals encounter in their habitats, such as average tree widths, rock sizes, and small mounds that they interact with on a regular basis.
2. As stated before, generally speaking waves having wavelengths shorter than the objects they interact with will be reflected, and waves with longer wavelengths than those objects will be diffracted rather than reflected.
3. Use 340 m/s for the speed of sound in air as an approximation, the audible frequency range as (20–20,000 Hz), and choose an infrasound frequency of 15 Hz to determine the wavelength of the waves that could be used by one of the two groups of animals chosen (elephants and bats). Based on your answer for the wavelength, which one of the groups do you conclude would use these waves, and for what purpose is the sound used?
4. Now choose an ultrasonic frequency of 50,000 Hz to determine the wavelength of the waves that could be used by the other chosen group of animals. Which one of the groups do you conclude would use these waves, and for what purpose is the sound used?
5. Reflect on your results by writing a short paragraph describing how the two groups of animals would specifically go about using such waves.

References

1. Young, Thomas (1804). "Bakerian Lecture: Experiments and calculations relative to physical optics". *Philosophical Transactions of the Royal Society* **94**: 1–16.
2. Nöldeke, C. (1990). "Compact disc diffraction" *The Physics Teacher* **28**, 484 (1990); doi: 10.1119/1.2343118.

Chapter 7
Polarization

As in the social context, polarization implies a preferred orientation or direction, or even a partial perspective on an originally impartial one. The effect of polarization is to alter a pre-existing condition that shows no preference for the way things are, or that has no particular direction or way for objects to move. This is essentially the way one can see waves as they would appear to oscillate in all allowed directions, and of course this has implications for the types of waves that can do that.

Exploratory Task
Use the simulation available at https://phet.colorado.edu/en/simulation/legacy/radio-waves
Make sure the screen looks like the figure below

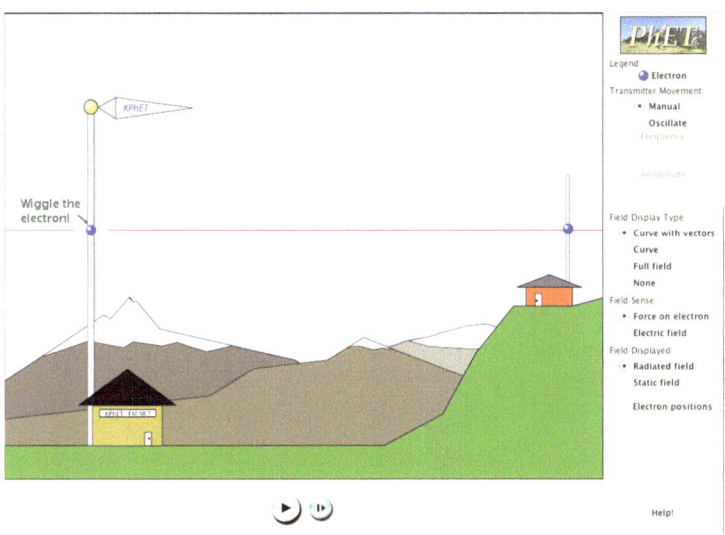

(continued)

© Springer International Publishing Switzerland 2017
F. Espinoza, *Wave Motion as Inquiry*, DOI 10.1007/978-3-319-45758-1_7

Now click on Oscillate and describe what you see after a few seconds; what reasons can you offer for what you observe happening?

Transverse waves result from oscillations that are perpendicular to the direction the wave travels whereas longitudinal waves are by definition already constrained to move in the way the oscillations occur. An un-polarized or non-polarized wave would exhibit the property of oscillations in every direction allowed, and the filtering effect, which is the result of polarization, would decrease the amount of oscillations in certain directions. Figures 7.1 and 7.2 demonstrate how polarization occurs.

Figure 7.3 demonstrates why longitudinal waves like sound cannot be polarized; the oscillations being along the same direction as the wave travel allow the wave to be transmitted regardless of the orientation of the slits.

There are innumerable uses of polarization with electromagnetic waves, so we shall concentrate on the applications that concern light.

Figure 7.4 shows the reason why electromagnetic waves can be polarized. In reality, the oscillations should occur in all directions, so the actual shape of an un-polarized

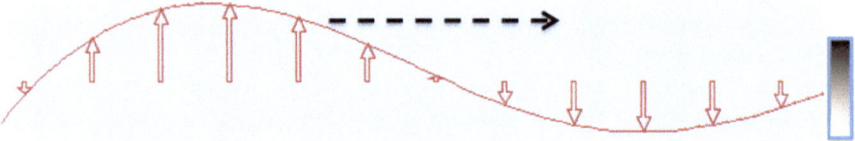

Fig. 7.1 A transverse wave moving to the right as represented by the *dashed arrow*; the *other arrows* illustrate the oscillations. The slit on the right will allow the passage of the wave since the orientation of its opening is along the same direction as the oscillations, namely along the vertical direction

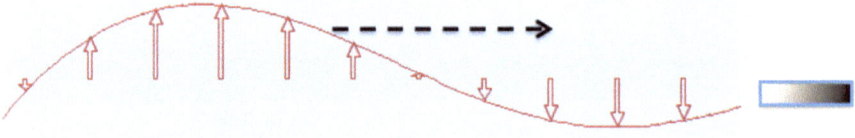

Fig. 7.2 The same wave will be blocked by the orientation of the slit's opening being horizontal, whereas the oscillations are vertical

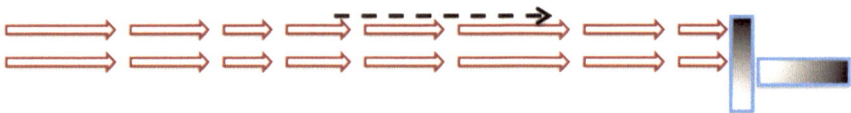

Fig. 7.3 The *arrows* now represent oscillations along the same direction as the wave travel, a longitudinal wave; in this case they will all go through regardless of the orientation of the slits

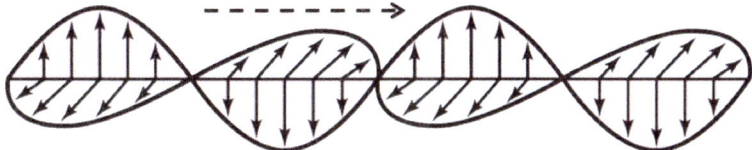

Fig. 7.4 An electromagnetic wave is clearly seen as consisting of oscillations of electric and magnetic fields along mutually perpendicular planes. One can assume the electric field oscillations to be represented by the *vertical arrows*, and those of the magnetic field to be represented by the horizontal ones, those that appear to be going into and coming out of the page. The *dashed arrow* represents the direction of travel of the wave

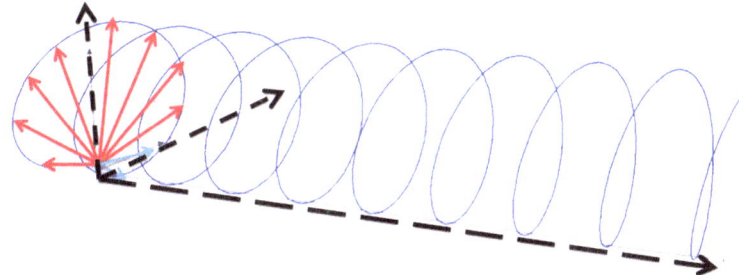

Fig. 7.5 An un-polarized or non-polarized transverse wave is represented by the helical pattern above, where the *two shorter dashed arrows* point along the plane of the oscillations, whereas the *long dashed arrow points* along the direction of propagation of the wave. The set of *solid arrows* drawn on the first circle are part of an infinite number of oscillations covering the plane, and they would point in every direction on it

electromagnetic wave should be visualized as a series of connected footballs. Polarization can be accomplished in several different ways: by absorption, by reflection, by double refraction, and by scattering. In each case, the number of oscillations on a plane perpendicular to the direction of wave travel is reduced. Figure 7.5 shows the details of a non-polarized wave.

Polarization by Absorption

Polarization by absorption consists of using filters made of materials whose molecules have been stretched along a chosen direction, and then allowing un-polarized waves to go through such filters. The orientation of the filter determines the amount of filtering or polarization, from a minimum to a maximum amount of blocked oscillations in a repetitive way. You can visualize this effect by looking at Fig. 7.1 and imagining the slit gradually rotated from its position in that figure, to the one in Fig. 7.2. If you were to continue rotating the slit until it returned to its original position, you would see the reverse of what happened before, namely an increase from a minimum amount of transmitted oscillations to a maximum.

Polarization by Reflection

Polarization by reflection can be illustrated by Fig. 7.6.

Figure 7.6 shows polarization by reflection as occurring whenever an un-polarized wave is incident upon a regular surface, and partial transmission and reflection take place. The angle of incidence and that of reflection are equal, thus ensuring complete polarization along the direction of the plane of the surface. The angle of refraction is smaller, although partial polarization in all directions takes place. The number of arrows is used to illustrate the amount of polarization, besides their orientation.

Polarization by Double Refraction

Polarization by double refraction is shown in Fig. 7.7.

Figure 7.7 shows double refraction leading to polarization. The originally un-polarized wave is incident upon a transparent material that has two different indices of refraction. Since the transmitted waves travel at different speeds due to the different indices, they emerge polarized and producing a double image. Notice that the two emerging waves have a polarization direction that is mutually perpendicular.

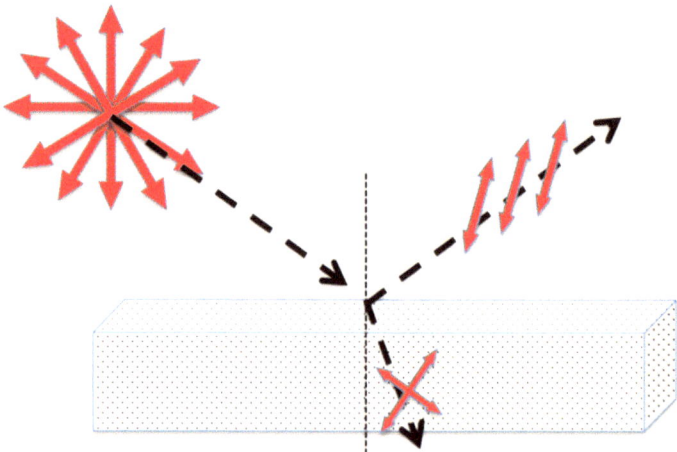

Fig. 7.6 Polarization by reflection occurs when an un-polarized wave is incident upon a surface where both transmission and reflection take place. The reflected wave is completely polarized along the direction of the plane of the surface, whereas the transmitted one is partially polarized in the other directions

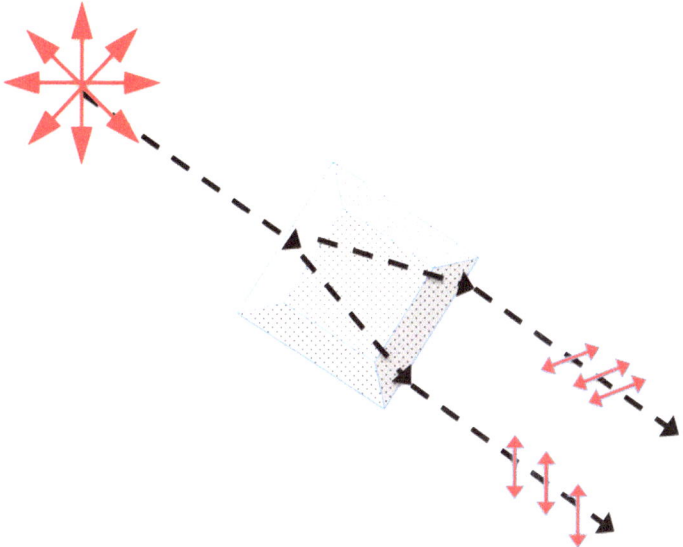

Fig. 7.7 Polarization by double refraction occurs whenever an un-polarized wave is incident upon and goes through a material with two different indices of refraction. These materials are called *birefringent*; since the first two refracted waves travel in different directions, the emerging waves will be mutually perpendicularly polarized and will produce a double image

Polarization by Scattering

Finally polarization by scattering results from un-polarized light being incident upon a material made of systems of particles, like clouds, where the electrons absorb and reradiate the light. The most important instance of polarization by scattering accounts for the different colors of the sky, and in particular why it is blue, as shown in Fig. 7.8.

When the electric field component of sunlight interacts with an atom or a molecule of air, it sets it in oscillatory motion with a resonant frequency. This oscillatory motion leads to a type of radiation like that coming from an antenna, and it is along a preferred direction, which is the polarization effect.

Figures 7.9 and 7.10 explain the phenomenon of a blue sky, and the appearance of the sun at dusk, when it is low on the horizon. In Fig. 7.9 the scattering that results in the blue color is shown as an interaction of an originally un-polarized electromagnetic wave (sunlight) with a molecule in the atmosphere. In Fig. 7.10 the reddening of the sky is shown from the perspective of observers on the Earth. Two observers are at points A and B, respectively, and they will see two different colors of the sky. The difference in the type of scattering depends on the wavelength of the light and the size of the particles that scatter it. The colors one observes in the sky depend on scattering; recall that the range of wavelengths for visible light is approximately (400–750) nanometers (nm). When a deep blue sky is seen, that comes from

Fig. 7.8 The figure shows two examples of polarization by scattering: the *blue color* of the background sky, and the *white color* of the clouds in the foreground

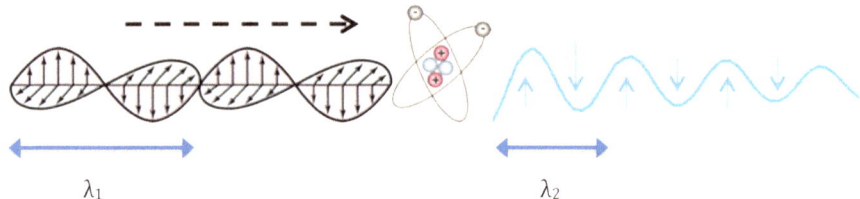

λ_1 λ_2

Fig. 7.9 The scattering of an un-polarized electromagnetic wave (the one on the left) by an air molecule can be shown by assuming that photons are oscillating in two directions. The oscillations of the electric field are along the vertical direction; the original wave has a wavelength λ_1, and upon striking the molecule the electrons will scatter the shorter wavelength photons producing a polarized wave of shorter wavelength λ_2

preferential scattering of photons of shorter wavelength in the visible spectrum. Particles larger than 750 nm scatter all colors equally, such as those making up clouds (1000–100,000) nanometers.

The reason why one sees a reddish sky when the sun is near the horizon is that the light waves are traveling a longer distance through the atmosphere, and they have already been scattered toward the shorter wavelengths. An observer will see the scattering resulting now from the longer wavelengths of the light spectrum, which is the red end.

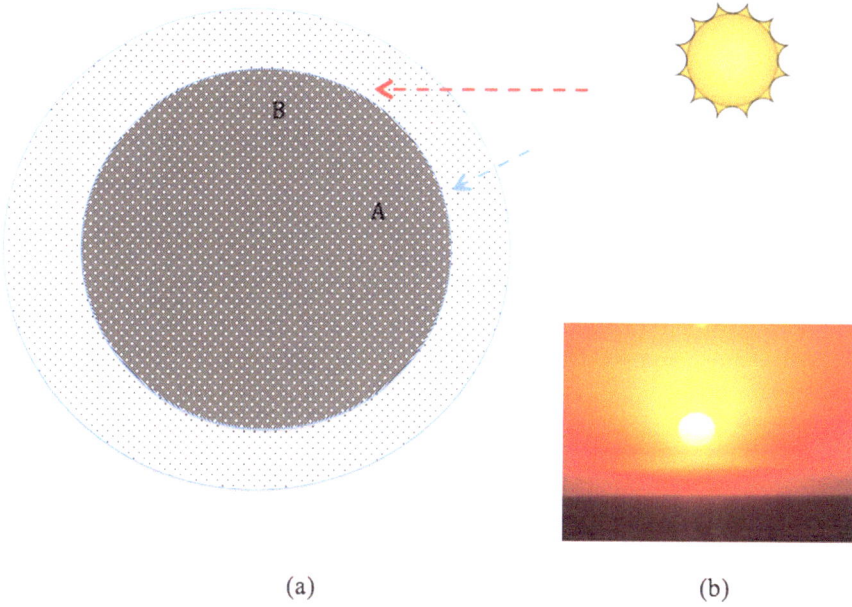

(a)   (b)

Fig. 7.10 Description of the observation of the color the sun appears when it is low on the horizon. The sun is drawn as it would look at a very long distance from the Earth, and so the diagram is not to scale. In part (**a**) an observer at A sees a blue sky, whereas one at B will see a reddish one due to two factors: the sunlight travels a longer distance through the atmosphere, and the blue color has already been scattered, thus creating the familiar view at dusk (**b**)

Conceptual Task

In order to get an idea of the scale needed in the next experimental task, consider that the visible range of the electromagnetic spectrum is roughly 400–740 nm.

$1 \text{ nm} = 1.0 \times 10^{-9} \text{ m}$

- When an object is said to be 10 micrometers (µm) in size, and knowing that $1 \text{ µm} = 1.0 \times 10^{-6} \text{ m}$

    (a) How big is the object in nm?
    (b) How many times larger than the object is the average wavelength in the visible spectrum?

- When an object is said to be 0.01 µm in size

    (c) How big is the object in nm?
    (d) How many times smaller than the object is the average wavelength from b)?

(continued)

Exploratory Task

(The following activity is based on the educational types provided by NASA (http://www.nasa.gov/centers/goddard/education/index.html)

Scattering of Light and Particle Size Determination

As previously discussed in the chapter on diffraction, light of certain wavelengths can either be reflected or diffracted, depending on the size of the objects with which the light interacts. If the wavelength is longer than the dimensions of the object, the light will be diffracted (provided there is no significant difference in the scale of the dimensions and the wavelength in question). If the wavelength is shorter than the dimensions of the object, the light will be reflected.

In the case of scattering, whenever the wavelength of the light interacting with the objects that scatter it is kept constant, then a similar situation to that of the previous paragraph takes place. In this instance, for a constant wavelength the size of the objects matters.

Given the range of wavelengths for the visible part of the electromagnetic spectrum where light is used to explore nature (400–750 nm), there is also a range of sizes for objects to either reflect or scatter light. Objects larger than about 10 μm reflect light, and those about 1/100 μm in size scatter light in all directions. However, equal amounts of light are scattered back toward the source and away from the source, and lesser amounts of light are scattered in other directions. Objects about 1 μm in size can exhibit strong forward scattering and weak backscattering.

- **Materials needed:**

  1. Laser pointer.
  2. Two clear plastic or glass water bottles or cups having vertical sides, not slanted, and having a range of 5–10 cm in diameter.
  3. Water, milk (1/20 teaspoon per 12 oz of water), and flour (less than or equal to 1/4 teaspoon—a "pinch"—per 12 oz of water).
  4. Eye dropper.
  5. A rotating platform or other means to rotate a container.
  6. Masking, duct, or electrical tape

- **Procedure**

  Attach a piece of tape to one side of each container. Fill one container with water and place it on the rotating platform. In the other container, prepare a highly dilute solution of milk, thoroughly mixed so the water is just slightly whitened. (Start with 1/20 teaspoon of milk per 12 oz of water; find the right proportions by experimentation in advance.) Place the laser and sample bottle on the rotating platform.

  Align the laser pointer so that the beam passes through the water bottle and projects onto the piece of tape on the far side of the container. (The tape will

(continued)

ensure that the laser beam is not projected farther into the room and perhaps into someone's eyes.) This arrangement is similar to the wavelength determination in chap. 6.

Darken the room and project the laser beam through the container of plain water. Observe the brightness of the beam in the water as the platform is rotated. The beam should pass straight through and be invisible or nearly so from all directions except directly along the beam.

Next, project the beam through the dilute milk solution. Laser light scattering from tiny particles of milk will delineate the laser beam. The intensity of the beam is stronger or weaker according to the scattering properties of the milk particles (primarily their size) as the assembly is turned in front of fixed observers. Observers should note how the beam reaches maximum brightness when they are looking in nearly the direction it is coming from.

Mix flour with the plain water (less than or equal to 1/4 tea-spoon flour per 12 oz of water) in the first container.

Project the laser beam through the dilute flour solution. The scattering properties of the milk and flour solutions are different because there is greater variation in flour particle size than in milk particle size. Store-bought milk is homogenized (its particles are reduced to the same size) so the cream stays in solution. With either mixture, notice how the beam intensity diminishes with distance (looking from the side).

You may have noticed when you drive under certain conditions that bright headlights in fog may or may not help drivers, depending on particle size. Reflections from large fog droplets make night visibility with bright headlights poorer than with dimmed headlights.

Predictions Based on Results

Try other materials that will remain suspended in liquid for sufficient amounts of time to allow for observations. First fill out the following table and predict whether or not the material will scatter light and thus be useful for particle size determination.

Material	Prediction	Observed
Cornmeal		
Cornstarch		
Oat bran		
Glitter		
Salt		
Sugar		
Sprite		
Diet Soda		

Reflections:

(continued)

Experimental Task

Polarized waves are those (transverse) where the waves vibrate in a specific direction. Polarization can be achieved by filtering some of the vibrations, such that only those aligned with the direction of the filtering mechanism will pass through. For the case of light waves, those waves that vibrate in the same plane as the polarizing material can pass through.

The most common method of polarization involves the use of a **Polaroid filter**. Polaroid filters are made of a special material that is capable of blocking one of the two planes of vibration of an electromagnetic wave. (Remember, the notion of two planes or directions of vibration is merely a simplification that helps us to visualize the wavelike nature of the electromagnetic wave.) In this sense, a Polaroid serves as a device that filters out one-half of the vibrations upon transmission of the light through the filter. When un-polarized light is transmitted through a Polaroid filter, it emerges with one-half the intensity and with vibrations in a single plane; it emerges as polarized light.

Light Sensor   Polaroid Filters   Light Source

Experimental Arrangement. Two Polaroid filters are used between a light source and a light sensor. The Polaroid filter closest to the light source is called the *Polarizer*, and the one closest to the light sensor is called the *Analyzer*. The objective is to keep the *Analyzer* fixed, and rotate the *Polarizer* a complete 360° to measure the transmitted light intensity. The light intensity is recorded by connecting the sensor to a device (computer or interface) where the information can be recorded.

(continued)

Detail procedure for turning the *Polarizer* while keeping the *Analyzer* fixed.

The transmitted light is linearly polarized by the Analyzer, and then its intensity is plotted as relationships between the light intensity and the angle of the Analyzer, as well as the time taken to turn it.

Procedure

1. We first determine the difference in light intensity when the sensor points directly at the light source, and then when a Polaroid filter is placed in front of the light sensor. Let's test both filters to ascertain whether or not they are identical in their polarizing properties.

Filter	Intensity without it (I_1)	Intensity with it (I_2)	Ratio (I_2/I_1)	% [$(I_2)/(I_1)$] × 100	% Difference from (0.50%)
1					
2					
Average					

2. Place the two Polaroid filters as shown in the first figure; while keeping the Analyzer fixed, turn the polarizer as shown in the detailed procedure and record your observations. Record the intensity for every 15° change, every 5 s until you complete one whole rotation or 360°.

(continued)

3. Plot two graphs of the intensity, first as a function of the angle, and then as a function of the time.

Observations (For Both Parts)
Processing The Data

1. Describe the graphs obtained; do they produce a wave pattern?
2. For what values of the degree measure was the light intensity a minimum and a maximum?
3. Summarize what you consider the sources of error in the experiment.

Chapter 8
Changes in Properties of Waves

The Doppler Effect

We learned in chapter two that the frequency of a wave is not a property of the motion of its components as the wave travels. As waves move through various materials their speeds, amplitudes, and wavelengths can change, but not their frequencies and correspondingly their periods. These are strictly properties of the manner in which the waves are generated. In other words, the frequency of a wave is constant, unless it is changed at its source. As with most statements in science, there are exceptions. An important one is the so-called Doppler effect (named after its discoverer Christian Doppler), an *apparent* change in frequency produced by motion. The use of the term apparent is meant to highlight the fact that it is relative to something. In this case the motion of either the source or the observer will cause a perceived change in frequency; however, the change isn't always there, it is only an effect produced by motion. Figure 8.1 shows the Doppler shift for sound waves.

As we shall see in this chapter, the apparent change in the frequency of signals generated by a moving object can be used for a great number of applications. Ordinarily speaking, to determine the average speed of a moving object one needs to know both the distance traveled and the time taken to travel it. The use of the term average implies an approximation since the object's speed may change and so determining its instantaneous speed proves more challenging. Of course, if the speed of the object is constant, the average and instantaneous values are identical.

Using the Doppler shift in frequency can help to determine a good approximation to the instantaneous value of the speed of a moving object, or at the very least a more accurate value than just the average.

The usefulness of the Doppler effect lies in its many applications to a variety of situations, where knowing the shift in frequency enables one to determine the

---

The original version of this chapter was revised. An erratum to this chapter can be found at
http://dx.doi.org/10.1007/978-3-319-45758-1_13

© Springer International Publishing Switzerland 2017
F. Espinoza, *Wave Motion as Inquiry*, DOI 10.1007/978-3-319-45758-1_8

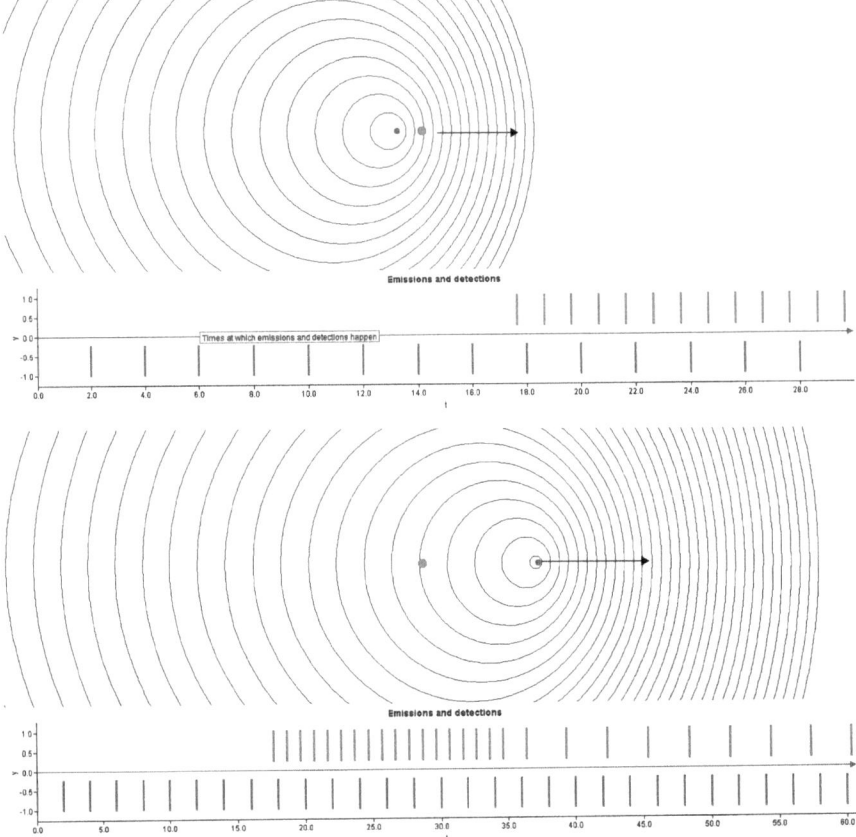

Fig. 8.1 Doppler shift of sound waves emitted by a source indicated by the *arrow*, as it moves to the right. As the waves reach the observer indicated by the *dot* to the right of the source, the spacing of the lines demonstrates that the frequencies are unequal. In the *top part* the observer measures a higher frequency (lines are closer to each other) as the source approaches it. In the *bottom part* the frequency measured by the observer is initially higher, and then lower after the source has passed it (the lines are more closely spaced and then further apart). However, the frequency of the source remains constant (lower scale in both parts)

speed of the object that produced the waves. This is true of such different situations as the flow of blood in the human body in physiological and anatomical studies, the motion of a speeding automobile in forensic investigations, and the motion of an air mass in meteorological forecasts. In order to appreciate the applicability of the Doppler effect, we need to look at the quantitative determination of the values involved in the variety of situations where the shift in frequency provides information about the motion of the objects in question.

We can state the general relationship between the frequency of a source of waves (f_s) and the observed frequency (f_o) when detected as either the source or an observer moves. The relationship takes into account the fact that when an observer moves either towards or away from the source, the speed of the waves relative to the observer is either larger or smaller than the original one in the equation $v = \lambda f$. Correspondingly, when the source moves either towards or away from an observer, the wavelength of the waves changes in a similar way to the speed for a moving observer. In other words, when there is no motion $f = \dfrac{v}{\lambda}$ from the above relationship. However, when there is motion the observed frequency is $f_0 = \dfrac{\Delta v}{\Delta \lambda}$.

When one combines both the motion of the source or that of the observer, the expression for the observed frequency becomes $f_0 = \left(\dfrac{V + V_0}{V - V_s} \right) f_s$.

In the above equation v is the speed of the waves, V_o is the speed of the object, and V_s is the speed of the source. The convention is to use positive (+) values for both V_o and V_s when motion is towards each other, and negative (−) when motion is away from one another.

Worked Example

An ambulance passes you by and you can measure the frequency as 560 Hz; if the actual frequency of the siren is 500 Hz, and you are stationary, how fast is the ambulance moving? We use 343 m/s as the speed of sound in air, and the Doppler formula states that

$$f_0 = \left(\frac{V + V_0}{V - V_s} \right) f_s.$$

$f_0 = 560$ Hz, $f_s = 500$ Hz, $V_0 = 0$ m/s, solving for V_s (the speed of the source, the ambulance)

Rearranging the equation above we get

$$\left(V - V_s \right) f_0 = V f_s$$

Expanding the term $V f_0 - V_s f_0 = V f_s$
    And solving for $V_s = V (f_0 - f_s)/f_0$
    Substituting the given values
    $V_s = (343$ m/s$) (560$ Hz-500 Hz$)/ 560$ Hz $= 36.8$ m/s

Example
A bat, flying at 5.00 m/s, emits a chirp at 40.0 kHz. If this sound pulse is reflected by a wall, what is the frequency of the echo received by the bat? ($v_{sound} = 340$ m/s.)
    ANS: 41.2 kHz

Other examples using the Doppler formula can be worked out by going to the *Explore Learning Gizmo* website
https://www.explorelearning.com/index.cfm?method=cResource.dspResourceExplorer&browse=Science/Grade+9-12/Physics/Sound.

You can use the formula to substitute the values given for the source frequency, the speed of the source, and the speed of sound. Note that the speed of the observer is zero in the formula, since the icon where the observed frequency is measured corresponds to an observer at rest. Select the box for "observed frequency" and then run the simulation. Does the value of the observed frequency match your calculated one?

You can practice by changing the given values, using the formula, and then running the simulation to compare the values.

Virtual Activity
The Doppler Effect and the Sonic Cone

The table shows six cases describing the emission of sound waves from a moving source towards the point on the right. The speed of the waves is constant (340 m/s), and each trial shows on the third column the effect of changing the speed of the source. Every trial shows the waves as they are about to be received at the other point.

1. Describe what the diagrams indicate about the changes in the speed of the source.
2. When does a cone first appear?
3. What happens to the shape of the cone as you continue to increase the source's speed values?

Trial	Speed of source (m/s)	Result
1	170	

(continued)

Trial	Speed of source (m/s)	Result
2	300	
3	340	
4	400	

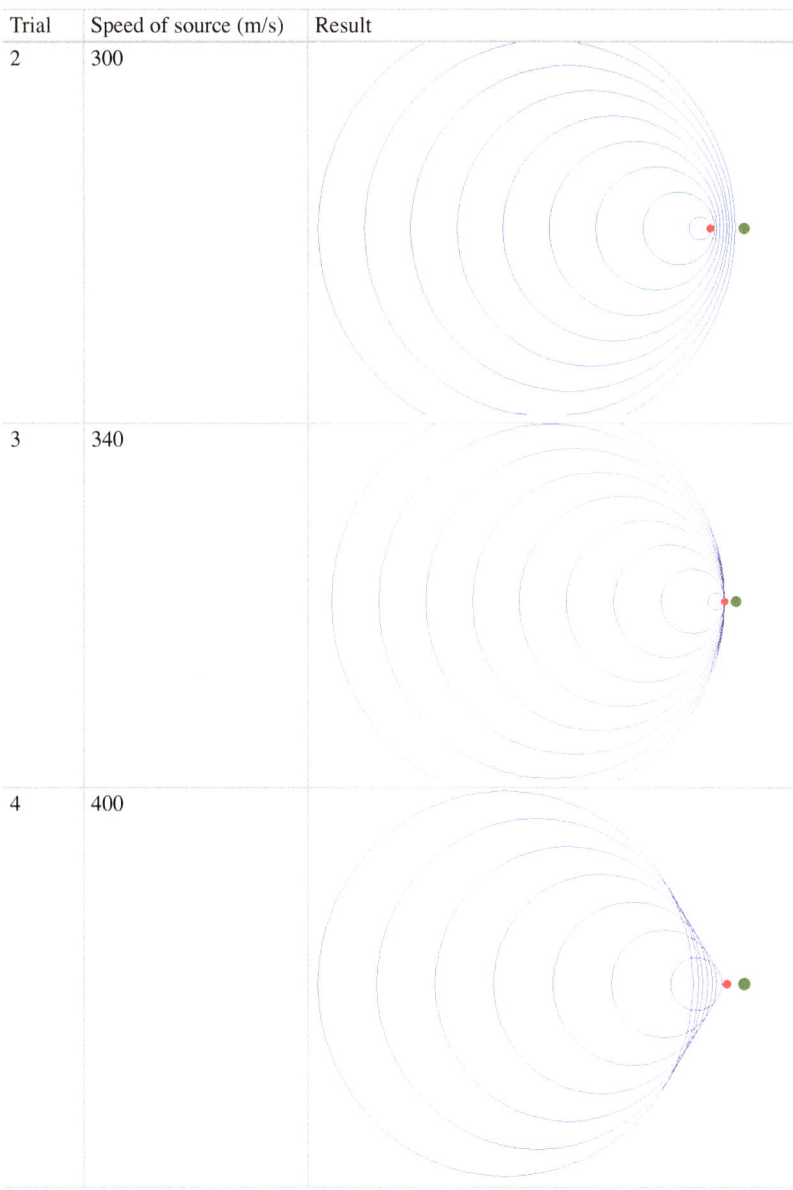

(continued)

Trial	Speed of source (m/s)	Result
5	500	
6	680	
6	800	

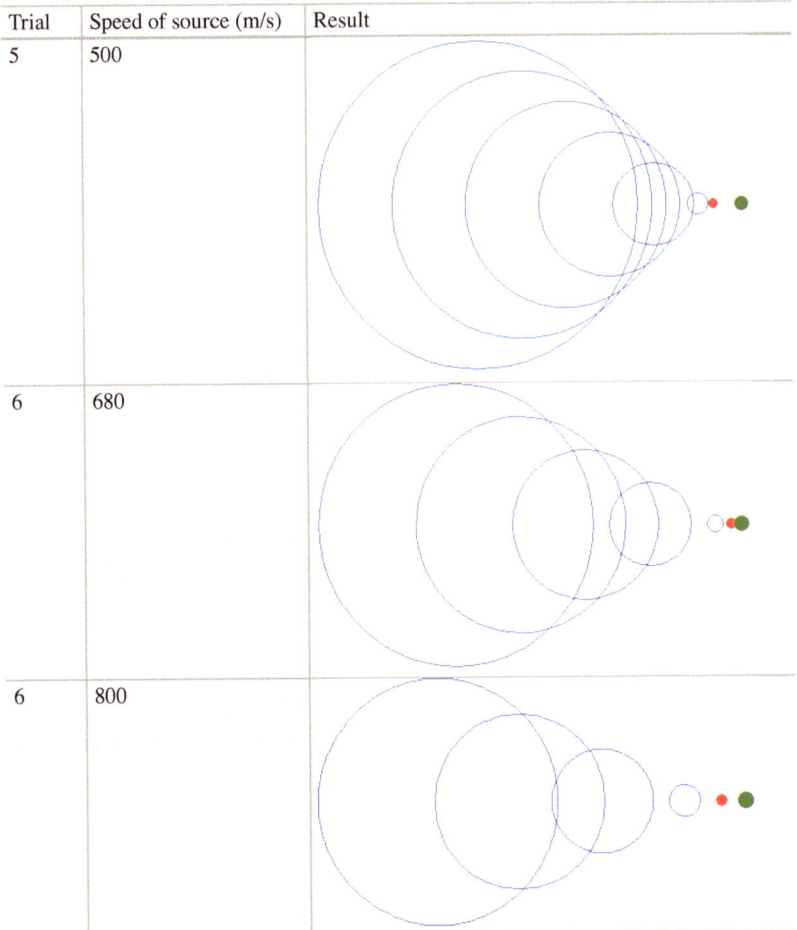

Exploratory Task
Using the Doppler effect to determine experimentally the Speed of a Moving Object

The speed of a moving object can be determined in various ways; the simplest case would be when the object moves with a constant speed. In general, when an object travels a distance D in a time t its speed is given by the formula

$$\text{Speed} = \frac{\text{Distance}}{\text{time}}. \quad V = \frac{D}{t}$$

(continued)

We can construct an experimental task where an object moves with a constant speed by having it move on a circular path. In this case the distance is the circumference, and the time is the period of the motion.

Hence $V = \dfrac{2\pi r}{T}$, where r is the radius of the circle described by the motion of the object.

We can also determine the speed of a moving object by using the Doppler effect.

Doppler Effect Formula for General Cases

- The source and the observer could both be in motion; the observed frequency f_0 is given by the equation
- $f_0 = f_s \left(\dfrac{V + V_0}{V - V_s} \right)$
- where f_s is the frequency of the source, v is the speed of the signal, v_0 is the speed of the object, and v_s is the speed of the source.
- If either one is approaching we use positive values for v_0 and v_s.
- If either one is moving away (receding) from the other we use negative values for v_0 and v_s.

Background Example

An alert student stands beside the tracks as a train rolls slowly past. The student notes that the frequency of the train whistle is 480 Hz when the train is approaching, and 440 Hz when it is moving away (receding). Using these frequencies the train's speed can be calculated.

Using
$$f_0 = f_s \left(\dfrac{V + V_0}{V - V_s} \right)$$
$V_0 = 0$ (the observer is at rest)
$V_s = V_t$ (the speed of the source is that of the train)

$V = 340$ m/s for the speed of sound

Approaching: $480 \text{ Hz} = f_s \left(\dfrac{340 \text{ m/s}}{340 \text{ m/s} - V_t} \right)$.

Receding: $440 \text{ Hz} = f_s \left(\dfrac{340 \text{ m/s}}{340 \text{ m/s} + V_t} \right)$.

Dividing the first equation by the second:
$$\frac{480 \text{ Hz}}{440 \text{ Hz}} = \frac{f_s \left(\dfrac{340 \text{ m/s}}{340 \text{ m/s} - V_t} \right)}{f_s \left(\dfrac{340 \text{ m/s}}{340 \text{ m/s} + V_t} \right)} = 1.09 \text{ (The } f_s \text{ terms drop out)}$$

Inverting the right side of the equation
$$1.09 = \left(\frac{340 \text{ m/s}}{340 \text{m/s} - V_t} \right) \left(\frac{340 \text{m/s} + V_t}{340 \text{ m/s}} \right)$$

(continued)

$$1.09 = \left(\frac{340 \text{ m}/s + V_t}{340 \text{ m}/s - V_t}\right), \text{ cross-multiplying}$$

$(1.09)\ (340 \text{ m/s} - V_t) = 340 \text{ m/s} + V_t$

$370.6 \text{ m/s} - (1.09)\ V_t = 340 \text{ m/s} + V_t$, rearranging the equation

$370.6 \text{ m/s} - 340 \text{ m/s} = V_t + (1.09)\ V_t$

$V_t\ (1 + 1.09) = 30.6 \text{ m/s}$

$V_t\ (2.09) = 30.6 \text{ m/s}$

$V_t = \dfrac{30.6 \text{ m}/s}{2.09} = 14.6 \text{ m/s}$ is the speed of the train.

To practice using the Doppler formula before undertaking the experimental task, determine the speed of a moving object where the following data were collected. An object moving on a circular path emitted a beeping sound where the frequency varied between 3500 Hz and 3200 Hz. What was the speed of the moving object?

Procedure

(Students work in groups stationed at various points near the circular path of an object that emits a sound (like a beep). The object is held securely by a string and made to rotate by a person holding the string at the center of the circle.)

1. Determine the speed of the moving object by measuring the length of the string (the radius) and then measuring the time for a given number of rotations (10). Measure the time for three trials and then average the result. Construct a table showing the data collected.
2. Divide the average time for 10 rotations by 10 to get the period of each rotation.
3. Use the equation $V = \dfrac{2\pi r}{T}$ to get the speed of the moving object.
4. Determine the speed by using a microphone interfaced to a device, such as a Vernier LabQuest to find the frequency of the beeping sound that the rotating object produces. Three frequency values are needed:

   (A) The frequency of the stationary object. (This frequency is only needed to verify that the approaching and receding frequencies are sufficiently different.)
   (B) The frequency when the object is approaching the microphone.
   (C) The frequency when the object is moving away from the microphone. To ensure sufficient accuracy these frequencies should be measured three times each, and then averaged. Construct a table for the data collected.

5. Use the Doppler formula as in the background examples to determine the speed of the rotating object.
6. Discuss the results of both procedures, and include a reflections section with specific references to the various sources of error.

(continued)

Write a Report including the following sections:

(A) Objective
(B) Brief procedure
(C) Data, calculations, and results
(D) Reflections—Analysis and discussion of sources of error.

Application to Light

The Doppler shift can be shown as it occurs in the case of light; the emission spectrum of Hydrogen is shown in Fig. 8.2a as it appears in the laboratory (at rest). The same spectrum is shown as it appears when Hydrogen is detected in the spectrum of stars and galaxies.

Figure 8.2 shows what happens to the Hydrogen spectrum. The lines in part (a) for a cloud of Hydrogen at rest appear around 430 nm for the blue, around 490 nm for the teal or blue–green, and around 650 nm for the red. In part (b) when the

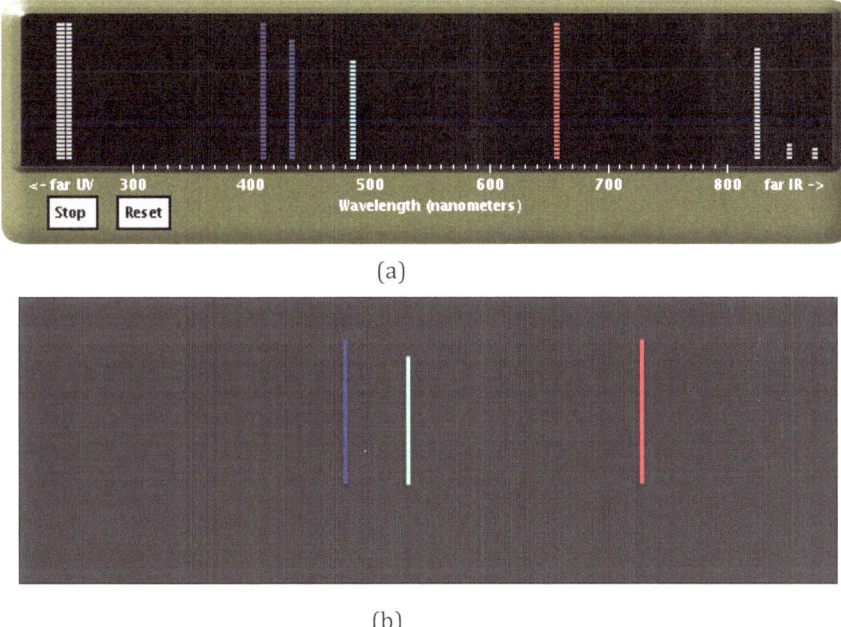

(a)

(b)

Fig. 8.2 The figure shows what happens in the case of light; the spectrum of Hydrogen gas observed in the laboratory shows lines of varying intensity. In part (**a**) the same lines are observed in the spectrum detected in a cloud of Hydrogen gas that is not moving with respect to an observer. In part (**b**) the cloud is moving away from the observer at a speed about 10 % of the speed of light. The spectral lines are shifted towards the longer wavelength or red end of the spectrum

spectrum is seen in a cloud that is moving and the same lines are now shifted to the right; the blue line now appears at around 480 nm, the teal line at around 540 nm, and the red at around 730 nm. The fact that they are all shifted towards the longer wavelength end of the spectrum is an indication that the cloud is moving away from the observer.

The longer wavelength end is the red part of the visible spectrum, hence the use of the term *redshift* in astronomy. It is one of the most valuable pieces of information available to astronomers and other scientists attempting to understand the large-scale properties of the universe.

Exploratory Task
Since radio signals are electromagnetic waves they can be used in radar applications; the Doppler imaging of clouds and air masses is used for weather prediction. Use an online source of information, such as The National Weather Service to write a brief explanation of how the Doppler effect is utilized in this case.

Human Hearing and the Subjective Perception of Changes in Sound Intensity

The intensity, I, of a wave is defined as the power per unit area. This is the rate at which the energy being transported by the wave transfers through a unit area perpendicular to the direction of the wave. A point source will emit sound waves equally in all directions, which can result in a spherical wave where the power will be distributed equally through the area of the sphere.

Figure 8.3 shows the structure of the human ear. The part that concerns us is the curled up membrane in the inner ear. It is called the basilar membrane and known to have a length of approximately 35 mm.

Place Theory of Hearing

The theory is based on the observation that two tones separated by a change in frequency that either doubles or halves the initial frequency (an *octave* interval such as 512 Hz/256 Hz$= 2/1 = 4/2 = 6/3 = 8/4$) corresponds to a length separation of about 3.5 mm along the basilar membrane, within the human audible range (20 Hz–20 KHz). Approximately ten such intervals covering the entire membrane have been identified, thus making the whole membrane (35 mm) equal to roughly ten octaves.

Figure 8.4 shows a comparison between two relationships. In part (a) the relationship is that between the magnitude of ground movement and energy released

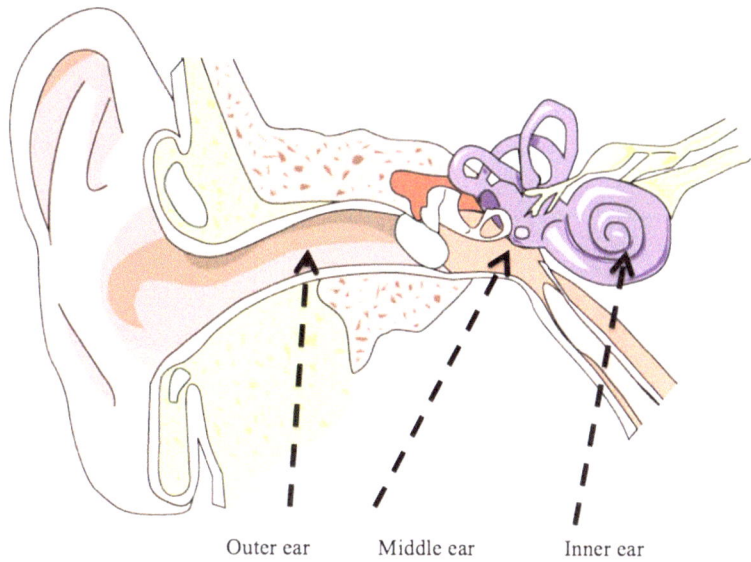

Outer ear   Middle ear   Inner ear

Fig. 8.3 The human ear can be divided into the outer, middle, and inner ear

during a seismic event. This relationship constitutes the basis for the Richter scale, which even in its modified version is used to determine the effect of earthquakes. Note that the growth is exponential, and the energy difference between a 1.5 and a 2.5 magnitude earthquake is about 300 units; whereas between a magnitude 2 and 3 earthquake the energy difference is 900 units. In other words, while the difference in magnitude is the same (1 unit), the energy difference is much greater between larger magnitudes. Thus an earthquake of magnitude 3 releases 900 times the energy of a magnitude 2 one, or it can be said to be 900 times stronger.

Part (b) shows a similar relationship between the frequency changes along the basilar membrane and the length of the membrane where there is a doubling of the frequency. Using the frequencies of several notes from the musical scale we get

$$C_3 - C_2 = 130.8\text{Hz} - 65.4\text{Hz} = 65.8\text{Hz};$$
$$C_4 - C_3 = 261.6\text{Hz} - 130.8\text{Hz} = 131.2\text{Hz};$$
$$C_5 - C_4 = 523.3\text{Hz} - 261.6\text{Hz} = 262\text{Hz}$$

$$C_6 - C_5 = 1046.5\text{Hz} - 523.3\text{Hz} = 523.3\text{Hz};$$
$$C_7 - C_6 = 2093\text{Hz} - 1046.5\text{Hz} = 1046.5\text{Hz}$$

The graph shows that for every doubling of the frequency the distance along the basilar membrane that responds to the change is roughly 3.5 mm. The distance in question is that between hair-like cellular structures that rise in the membrane and that lead to nerve responses correlated with frequency.

a

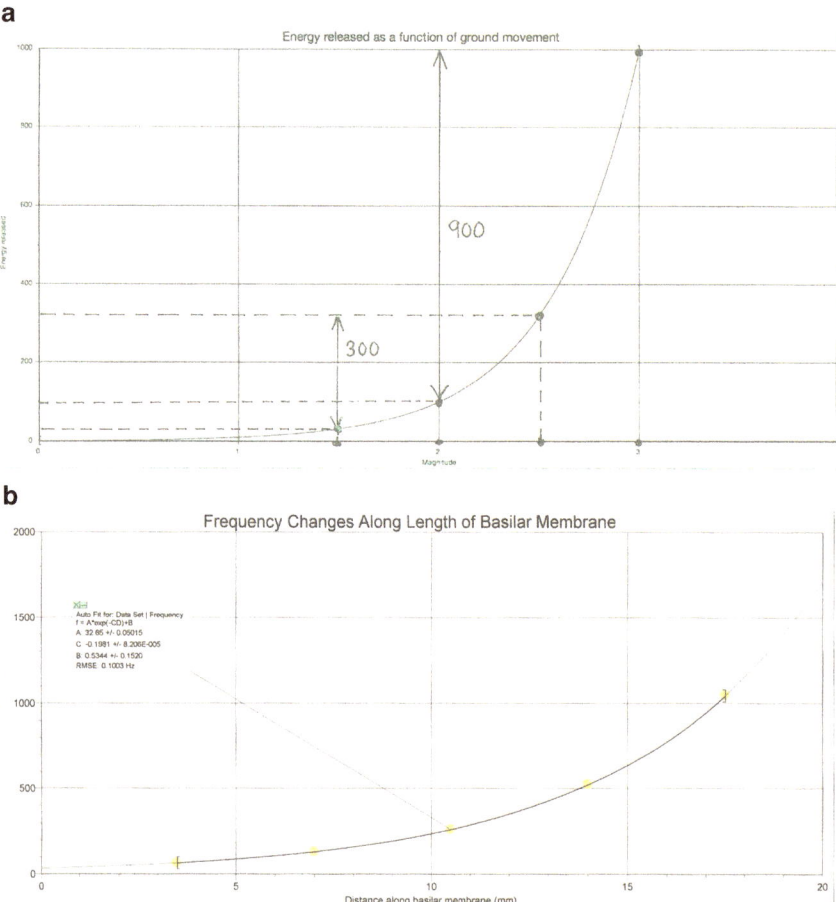

Fig. 8.4 Comparison of the scales used for both seismic energy release as a function of the magnitude of the ground movement (**a**) and the frequency response of the human ear as a function of the length along the basilar membrane (**b**)

The dependence of human responses to sound intensity changes based on frequency needs an expression as a function of this doubling of the frequency, so the relationship cannot be linear. Consequently, we can express as a *power* function the relationship between variables where one doubles constantly as the other one increases by a fixed amount. However, as you can see from Fig. 8.4a, a magnitude 7 earthquake would require extending the axes and it would be 10 million times stronger than a magnitude 1 earthquake. Clearly the scale of one of the axes needs to be reduced to display such large numbers in a comparable proportion to the values of the other. There is a relationship between power and logarithmic functions; they are the inverse of each other. In other words, the inverse function of $y = b^x$ is $y = \log_b x$. Therefore, a relationship expressed as a function that yields a set of values expressed as powers can also be expressed by a logarithmic function that expresses these values as regular numbers.

The human ear is known to respond to changes in air pressure where the intensity can be expressed in W/m² but the range between what is audible (called the threshold of hearing) and what is bearable (the threshold of pain) is about a trillion times, or 1.0×10^{12} times. To reduce such a huge range of values to a scale where the sound level can be expressed in a more manageable way, we can use an expression that involves a logarithmic function.

$$SL = (10dB)\log_{10}\left(\frac{I}{I_0}\right)$$

where I is the given value of the intensity, and I_0 is the threshold of hearing (the lowest possible sound) $= 1.0 \times 10^{-12}$ W/m²

It is imperative to clarify at this point that the use of SL as the sound level is meant to represent what we hear, not the actual energy content of the sounds, which in some textbooks is expressed in terms of loudness curves in addition to intensity values. The emphasis on the use of *sound level* for SL is to differentiate it from that usage in this context.

This human response can be represented by curves that are constant in intensity (energy content), but variable in sound level (hearing volume) depending on the frequency at which they are heard. The curves are based on the work of Fletcher and Munson at Bell labs in the 1930s, and were made by asking people to judge when pure tones of two different frequencies were equally loud, the curves being the average results from many subjects. Several of these curves are shown in Fig. 8.6;

Fig. 8.5 Representation showing that the ear canal acts as a 2.5 cm close-ended tube

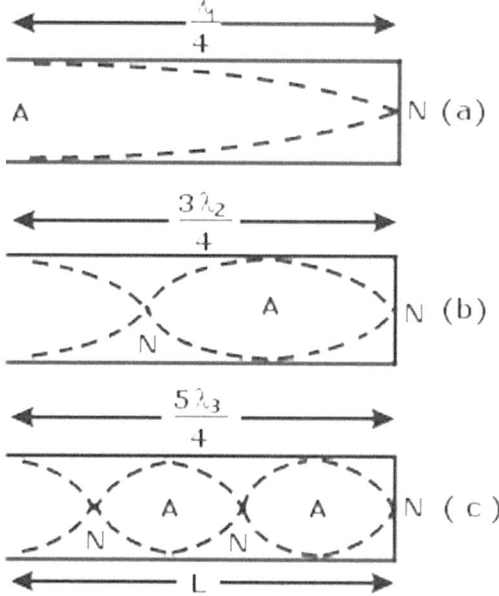

the lowest three beginning with the threshold of hearing, a medium one representing typically heard everyday sounds, and the highest one at the threshold of pain.

The lowest curves indicate that low energy sounds are not all heard equally at low frequencies, and the highest one is painful at almost all frequencies, represented by its increased flatness. The common dip of all the curves is a result of the shape of the ear canal.

Figure 8.5 shows a representation of the ear canal as a tube closed at one end. Note that the length of the canal is ¼ of a wavelength, or $L=\lambda/4 \rightarrow \lambda=4$ $L=4$ (0.025 m)=0.10 m

$$v=\lambda f \rightarrow f = \frac{v}{\lambda} = \frac{340\,m\,/\,s}{0.10\ m} = 3400\ Hz,\ \text{which is the fundamental frequency } f_1.$$

The next frequency is $f_3\ (3f_1)=3\ (3400\ Hz)=10{,}200\ Hz.$

Note that these values correspond to the dips in all the loudness curves shown. This is the result of two resonances seen as the anti-nodes (A), or maximum amplitudes. That is why the ear exhibits such sensitivity at those frequency values (3400 Hz and 10,200 Hz) as shown in Fig. 8.6.

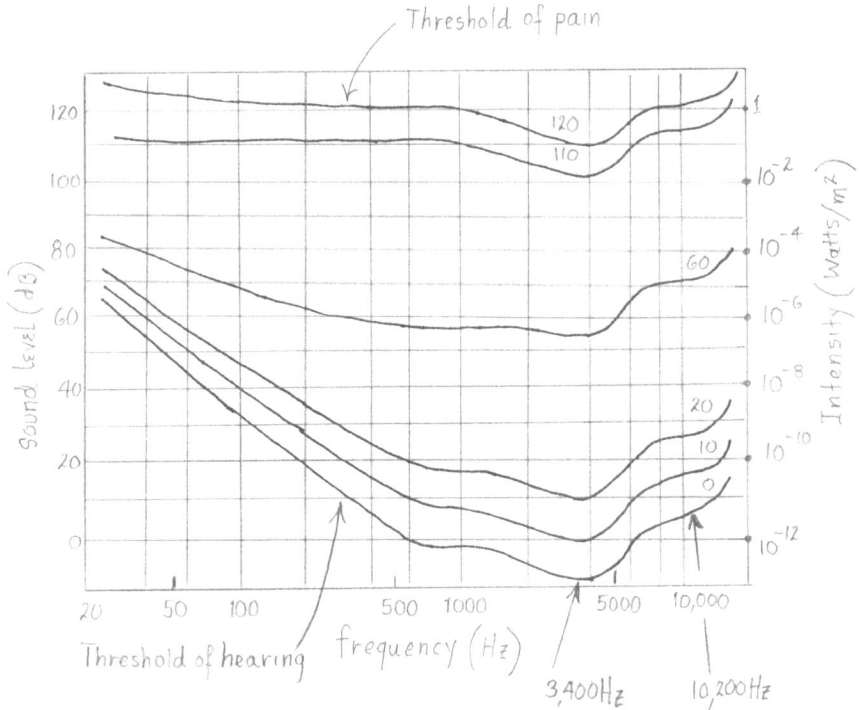

Fig. 8.6 Relationship between intensity levels and sound levels perceived as the frequency of the sounds changes. The entire range of audible intensities (one trillion times difference in W/m²) can be condensed into a manageable scale in dB by using a logarithmic relationship with the base of 10. The two resonant frequencies (3400 Hz and 10,200 Hz) where the ear is most sensitive are shown

Table 8.1 A comparison between ways to classify changes in sound energy. As changes in intensity, and as changes in sound level or volume

Sound intensity (W/m²)	Loudness level change (dB)	Volume loudness (how many times louder)
10,000	40	16
1000	30	8
100	20	4
10	**10**	**2.0**
4.0	6	1.52
2.0	3	1.23
1.0	**0**	**1.0**

Restating the formula $SL = (10 \text{ dB}) \log_{10} (\frac{I}{I_0})$ we can see from the above graph that when a given intensity is $I = 1.0 \times 10^{-12}$ W/m², the value of SL is 0 since $\log_{10}$ (1)=0. If the given intensity is 1, then $SL = (10 \text{ dB}) \log_{10} (1/1.0 \times 10^{-12}) = (10 \text{ dB}) \log_{10} (1.0 \times 10^{12}) = 120 \text{ dB}$, as can be determined from a logarithm calculator table or a calculator with logarithm functions. These two values correspond to those on the left vertical axis. The table below illustrates how the formula gives the corresponding values of SL for some given intensities, and how many times louder these values are compared to other levels, something that is not very intuitive on first inspection. As an example, consider that a decrease of 10 dB in SL is equivalent to half the volume but 1/10th the intensity.

Useful Relations

$\text{Log } ab = \log a + \log b$

$\quad \text{Log } a/b = \log a - \log b$

$\quad \text{If } y = A^x, \text{ then } x = \log_A y$

$\quad$ In other words, if $y = 10^x$, then $x = \log y$

As Table 8.1 shows for the values indicated, a doubling in sound level corresponds to an increase in sound intensity of ten times; this is true regardless of where on the decibel scale the values change. Suppose operating a portable vacuum cleaner produces a sound such that the sound level is measured at 60 dB, how much would the sound made by four identical machines be?

Solution

We first determine the intensity value of the sound at 60 dB, by using the formula

$$SL = (10 \text{ dB}) \log_{10} \left(\frac{I}{I_o} \right)$$

$60 \text{ dB} = (10 \text{ dB}) \log_{10} \left(\frac{I}{I_o} \right)$, dividing both sides by 10 dB

$6 = \log_{10} \left(\frac{I}{I_o} \right)$, using the last of the relations given above (if $x = \log y$, then $y = 10^x$)

$$\frac{I}{I_o} = 10^6 \rightarrow I = I_o \times 10^6$$

$I = (1.0 \times 10^{-12} \text{ W/m}^2) (10^6) = 1.0 \times 10^{-6} \text{ W/m}^2$

This is the intensity of one machine; four such machines will produce $4.0 \times 10^{-6} \text{ W/m}^2$

And so the new sound level value will be

$$SL = (10 \text{ dB}) \log_{10} \left(\frac{I}{I_o} \right) = (10 \text{ dB}) \log_{10} (4.0 \times 10^{-6} \text{ W/m}^2 / 1.0 \times 10^{-12} \text{ W/m}^2)$$

$$SL = (10 \text{ dB}) \log_{10} (4.0 \times 10^6) = (10 \text{ dB}) (6.6) = 66 \text{ dB}$$

A Note of Caution about Solving Sound Level Problems

The use of the two scales that are connected through the logarithm function, the intensity scale, and the sound level scale can be counterintuitive. When solving sound level problems, it is imperative that we understand the distinction that must be made between sound level differences, as opposed to sound intensity differences. In the previous example, the noise difference between one machine producing a sound level of 60 dB, and four such machines must be determined first in terms of the intensities. The real difference (the energy content) is between the intensities. The sound level of the four machines depends on what their collective intensity is; one may be tempted in the above example to simply multiply the 60 dB by 4, which would yield an unreasonable sound level (240 dB); this number would exceed the noise produced by a jet engine measured at a short distance from it. If we think about the result, would it be reasonable to expect that four portable vacuum cleaners would be noisier than a jet engine? (A look at a comparison chart of sound levels in decibels will convince anyone that a sound level higher than 200 dB isn't physically possible.)

Practice Problems

(1) In a workplace a machine has a sound level of 80 dB, how many identical ones can be added before exceeding the federal noise limit of 90 dB?

(2) If an orchestra produces a sound level of 85 dB, and a single violin produces 70 dB, how does the intensity of the sound of the orchestra compare to that of the violin?

(3) A busy street has 100 cars/min passing a given point during a weekday producing a sound level value of 70 dB; if the number is reduced to 25 cars/min during the weekend, what is the resulting sound level value?

Exploratory Task (I)

Noise exposure can be a serious matter if the levels approach high enough values on the decibel scale to cause discomfort, and there are specific regulations concerning indoor noise levels that humans can be exposed to before harmful effects appear. We can use a sound level meter, or download an App that enables a phone to collect sound level values and determine the level of exposure in a given space.

The following table can be used as a reference to compare the readings collected during a given period, with what has been determined to be acceptable indoor exposure.

Duration per day (h)	Sound level (dB)
8	90
6	92
4	95
3	97
2	100
1.5	102
1	105
0.5	110
0.25	115

US Dept. of Labor
Occupational Safety and Health Standards
1910.95(b)(2)
Table G-16.

Exploratory Task (II)

Determining the relevance of measurements of decibel levels to the housing market

Use the article "Soundproofing for New York Noise" (The New York Times, Real Estate section, December 11, 2015) to determine three specific uses of sound properties introduced in the text so far, that are actively used in trying to address noise issues faced by apartment building tenants.

References

1. Roederer, J. G. (2008) *"The Physics and Psychophysics of Music"* Fourth Edition. Springer.
2. Berg, R. E. & Stork, D.G. (2004) *"The Physics of Sound"* Third Edition; Pearson/Prentice Hall.
3. Bennett, J. et. al. (2007). *"The Essential Cosmic Perspective"* Fourth Edition; Pearson/ Addison-Wesley.
4. Serway, R. A. & Jewett, J. W. (2014). *"Physics for Scientists and Engineers"*. Ninth Edition; Cengage.

Chapter 9
Wave Propagation and Intensity Variations

All transfer of energy by waves obeys the same laws that can be stated for light and sound, particularly as it relates to the measurements of certain properties and the variables they depend on. The very first property is that of the inverse-square dependence on the distance from a source of waves. This is of the utmost importance due to the similarity exhibited by the behavior of masses, charges, and systems made of these. The most commonly observed phenomenon is that displayed by the electric, magnetic, and gravitational forces. They all depend on the inverse-square of the distance between the interacting objects. This has far reaching implications for the way all forms of energy as radiation spread out and interact with matter.

Exploratory Task

For this activity you need to download an App (Sound Level meter) to your phone so that you can measure the sound intensity produced by the online-generated signals. Use a source of sound that can register a measurable level of loudness on a sound level meter App on your cell phone (such as *Insta Decibel* for the iPhone).

To carry out the tasks go to

http://onlinetonegenerator.com/

Select the 432 Hz tab and then change the frequency to 440 Hz. Position your phone as close to the sound source as possible. Play the sound produced by each of the four different tones and measure the sound intensity value for each tone with the Sound Level meter App running on your phone. Rank the tones from loudest to softest; use the loudest sound and position the phone with the Sound Level meter App running at selected distances from the source of the sound (computer/laptop/tablet) and record the loudness at each position, so that you have at least six different values.

(continued)

© Springer International Publishing Switzerland 2017   167
F. Espinoza, *Wave Motion as Inquiry*, DOI 10.1007/978-3-319-45758-1_9

Begin as close to the source of sound as possible and then record the loudness level at each position of the phone as you move it away from the source.

Plot a graph of the loudness as a function of the distance from the source, and draw the best-fit curve that connects the points.

What sort of relationship does your graph follow between the sound intensity and the distance from the source?

Reflect briefly on the likely sources of error in this experimental activity.

There is a way to classify the different types of radiation that exist, similarly to the way we classified waves earlier on.

Experimental Task

Inverse-Square Dependence on Distance

Background

In this experiment we shall explore the relationship that exists between several properties of the physical universe and the distance between the objects that interact and that give rise to such properties. Among these are the inverse-square dependence on distance of the gravitational force between masses, and the dependence of light intensity at a surface on the distance from the source of light.

To begin please predict in two ways, as a statement and then graphically what the graph of light intensity measured at various distances from a source of light will look like as one moves away from the source. In this experiment we are using a light sensor to determine the light intensity as read by the sensor, as we move it away from the source of light.

I. Using a Light Source

Apparatus needed for the experiment (the light source may be different than the one shown but this will not change the setup). The activity can also be performed by using an App that measures light intensity (such as Lux Camera for the iPhone), and then placing the phone at respective distances from a light source. The data can then be collected and plotted the same way as described in this particular experiment.

(continued)

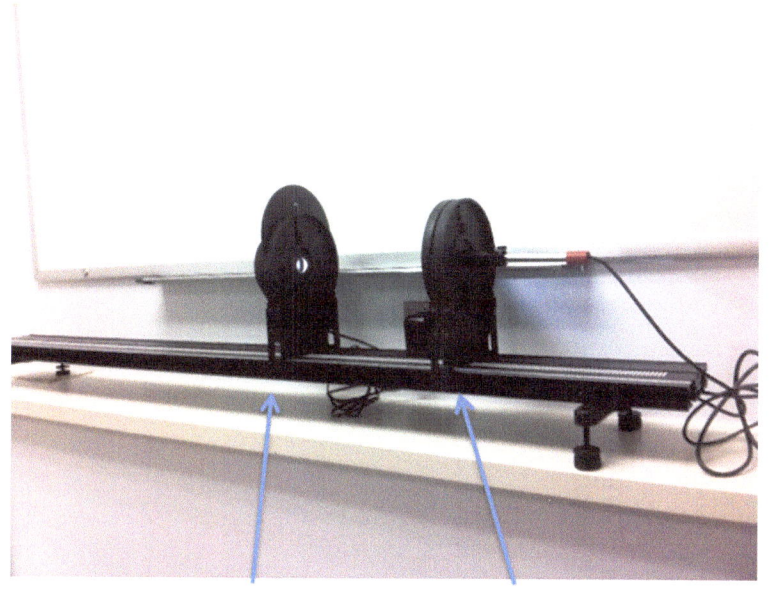

Light Source   Light Sensor

Prediction:

_____

_____

Predicted Graphical Relationship

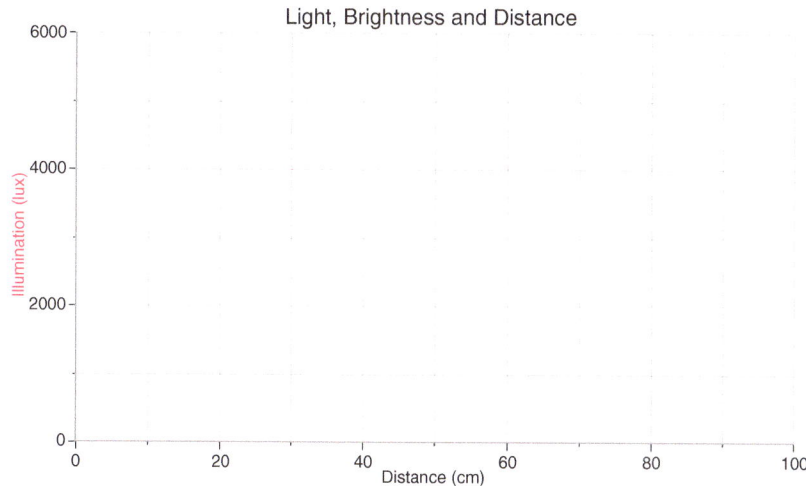

(continued)

Procedure

Now assemble the apparatus so that you can place the light sensor at several distances from the light source and then for each distance, take the reading of the light sensor and wait until the instrument measures the intensity for a few seconds and the reading stabilizes, then store the data and continue changing the distance by 10 cm for every trial. At the end of the trials determine the average intensity at the given distance and then fill in the table below.

Distance (cm)	Intensity (lumens)	Distance (cm)	Intensity (lumens)
10		50	
20		60	
30		70	
40		80	

Now plot the data on the graph below; connect the points with a best-fit curve, and then compare your obtained graph with your predictions.

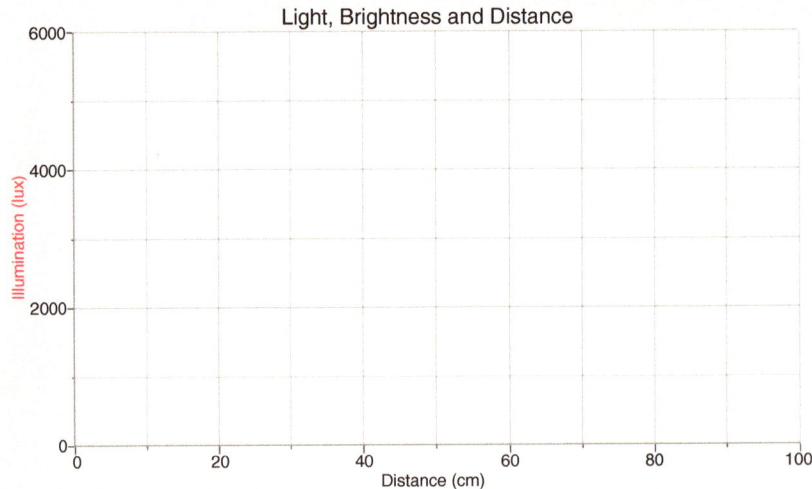

(continued)

Reflections on the Results:

II. Relationship Between Light, Sound, and Gravity

The equation for the gravitational force between two masses M_1 and M_2 separated by a distance d is given by

$$F_G = G M_1 M_2 / d^2$$

where G is the universal gravitational constant 6.11×10^{-11} (N) (m²)/(kg²).

The equation can be modified so that the changes in the distance between two masses M_1 and M_2 can be recorded and the corresponding force F' can be expressed in terms of F.

In the following table fill in the resulting value F' of the gravitational attraction between the objects M_1 and M_2 at the given distance, in terms of F. (Fill in the missing blanks for F'.)

$$F \approx M_1 M_2 / d^2$$

M_1	M_2	d	F'
1	1	1	F
2	1	1	$2F$
1	1	2	$F/4$
2	2	1	
2	2	2	
2	2	1/2	
3	2	1	
2	3	2	
3	3	3	
3	3	1/2	
4	3	2	
4	4	2	
4	4	4	
4	5	1/3	
5	5	1/5	

(continued)

Plot the data in the following graph and draw the curve that connects the points as a best fit.

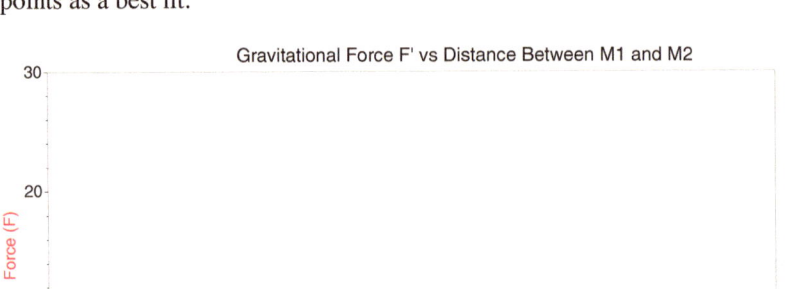

How does your graph of the gravitational force as a function of the distance between masses compare with those of Light Intensity vs distance, and Loudness vs distance?

The propagation of waves obeying the inverse-square dependence on the distance can also be demonstrated by a laboratory activity.

9.1 Radiation

At this point we shall introduce a topic that has loomed large throughout the text, but that we have not explicitly addressed despite having dealt with many of its properties, as we have explored those of waves.

Since we have been dealing with the way waves propagate from a source in this chapter, we can logically extend our discussion to what sources of light, sound, and other forms of energy do in general. We can begin with a definition of radiation as a way to generalize what we have explored so far in this chapter.

Radiation can be considered as the energy emitted from a source, and that travels through space or matter in the form of waves or high-speed particles. It should be pointed out that radiation is only one type of energy transfer; there is also conduction and convection as ways for energy to travel. However, the distinctive feature of radiation is that the way the energy is emitted is in all directions, as though it were emanating from a point.

1. Radiation by its types: electromagnetic (waves) and particle radiation.
2. Radiation by its effect: ionizing and non-ionizing.

As with our wave classification there is a crossover between the divisions, namely the first group can display both types of the second, but the reverse is somewhat more complicated. In other words, electromagnetic as well as particle radiation can both be either ionizing or non-ionizing. However, the distinction between waves and particles needs to allow for the difference in our perception of what particles are. There are components of electromagnetic radiation that behave like particles (photons), but they have properties that differ significantly from the particulate examples we encounter everyday.

As far as high-speed particles that are called particle radiation, they too can be ionizing or non-ionizing depending mainly on their speed and charge. Radioactivity (which incidentally also exhibits an inverse-square dependence on the distance from the source) is often described in terms of three types, as shown in Fig. 9.1:

(a) Alpha (a helium nucleus)
(b) Beta (fast-moving electrons)
(c) Gamma (high-energy photons)

Alpha particles are generally slow moving and not very energetic since they can be stopped by thick paper, or even by soft tissue. Beta particles can penetrate further than Alpha, but they lose energy upon colliding with atoms and larger objects. They can be stopped by a thin sheet of Aluminum, and even by human skin, although they can be harmful if swallowed. Gamma radiation is usually called rays since they are very energetic, and unlike Alpha and Beta radiation, have no charge. This is a reason why they can penetrate thick materials and can be harmful to human exposure, but can be stopped by lead or concrete.

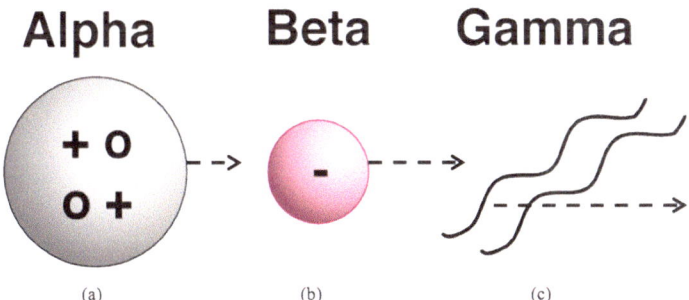

Fig. 9.1 The three types of radiation are illustrated as models of what they look like. In part (**a**) an Alpha particle is shown as a Helium nucleus, where the collective positive (+) charge is that of the protons, the neutrons are shown as (0) having no charge. In part (**b**) the beta particle is shown as a negatively (−) charged (electron) fast-moving one. In part (**c**) the Gamma ray is shown as a high-energy wave. The lengths of the *dashed arrows* are meant to illustrate the differences in speed, not to scale

We also need some clarification as to what the terms ionizing and non-ionizing stand for. According to our chemical models of molecular structure electrons are typically tightly bound to atoms through chemical bonding between molecules. This represents a case of chemical equilibrium where energy is mainly exchanged internally, although some is also given or taken from the environment. However, whenever an external source of energy radiates in such a way that the radiation has a certain amount of energy, it can disrupt the state of equilibrium that exists in matter. The following figure illustrates the case of ionization, when the chemical structure of matter is disrupted, creating a situation where there is now radiation being emitted to the surrounding environment. In this rather simplified explanation, the idea is that ionization leads to an increase in chemical activity, which can result in damage to organisms through a variety of means, particularly through mutations in biological activity.

Non-ionizing radiation by contrast can still cause increased activity between molecular structures, but without enough energy to cause the stripping of electrons, which is essentially what ionization is all about.

Figure 9.2 represents the basic idea behind ionization, where the resulting increase in chemical activity leads to a subsequent increase in biological activity that often results in harmful mutations to organisms. In this regard, ionizing radiation is to be avoided whenever one is exposed to either electromagnetic radiation, or any other type.

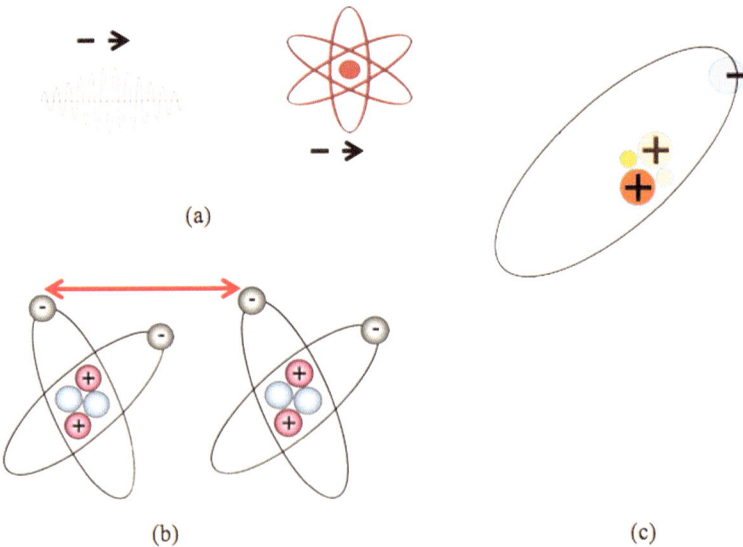

(a)

(b)   (c)

Fig. 9.2 A high-energy wave or a fast-moving particle impacts a molecule. In part (**a**) the wave is shown as a packet, and the particle is shown as oscillating in more than one dimension but moving along the *dashed arrow*. In part (**b**) the *solid arrow* represents the chemical bond between the atoms. If either the wave, or the particle possesses enough energy to break the bond, then ionization occurs. In part (**c**) the result of the collision is an ion, where the single electron in the orbit is unstable since the nucleus has a different amount of charge

Virtual Experiment

When light interacts with matter a number of different outcomes can be observed, depending on the energy of the light, which in turn depends on its frequency and wavelength, and the type of molecules the light interacts with.

Explore the various ways that light interacts with matter by using the PhEt simulation "Light and Molecules" available at

http://phet.colorado.edu/en/simulation/molecules-and-light

Run the simulation and explore the interaction of each type of radiation with a given molecule. Change the intensity of the radiation and observe what happens to the molecule, describing whether or not the chemical bonding is disrupted. Fill in the table below with your observations for each molecule.

	Microwave	Infrared	Visible Light	Ultraviolet
CO				
N_2				
O_2				
CO_2				
H_2O				
NO_2				
O_3				

Since radiation is contained in all parts of the electromagnetic spectrum, we need to display in what parts of it the distinction between ionizing (harmful) and non-ionizing radiation can be found.

Figure 9.3 contains the region of the spectrum where the radiation changes from non-ionizing to ionizing. The boundary cannot be made sharper than somewhere in the ultraviolet region since there is uncertainty as to where exactly one ceases to deal with strictly non-ionizing radiation, and then encounters the harmful effects of ionization. One can see that the visible part of the spectrum contains both types of radiation; however, the transition has taken place by the time one encounters the ultraviolet region.

Conceptual Task

A cell phone's SAR, or its Specific Absorption Rate, is a measure of the amount of radio frequency (RF) energy absorbed by the body when using the handset. All cell phones emit RF energy and the SAR varies by handset model.

Find your phone's SAR rating (value) and determine whether it falls on the low or high emission range according to the government's standards.

https://www.fcc.gov/encyclopedia/specific-absorption-rate-sar-cellular-telephones

THE ELECTROMAGNETIC SPECTRUM

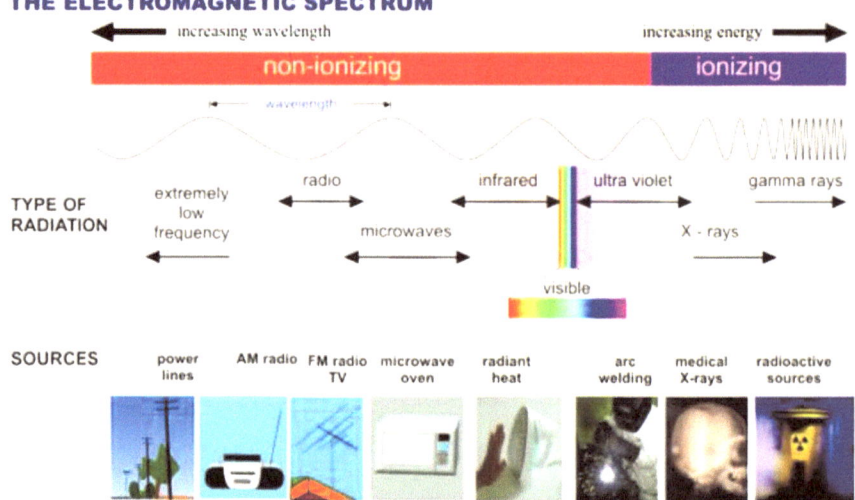

Australian Radiation Protection and Nuclear Safety Agency

Fig. 9.3 The figure illustrates the region of the electromagnetic spectrum where the transition between non-ionizing and ionizing radiation has taken place

Exploratory Task

The following frequency ranges are used in telecommunications:

Radio frequency band (30 KHz–300 GHz); cellular mobile [(872–960) (1710–1875) (1920–2170)] MHz; microwaves (2200 MHz–60 GHz).

According to the following figure of the electromagnetic spectrum, the cell phone band is in the non-ionizing region. However, there is concern in some circles about the effects of cell tower and cell phone exposure to radiation.

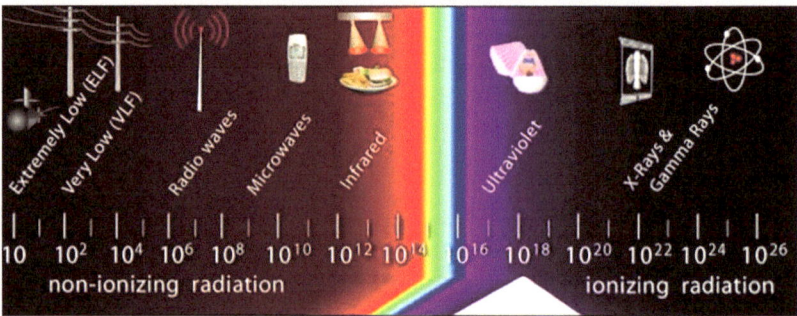

US Dept. of Labor-Occupational Safety and Health Administration

(continued)

- Have you ever read the fine print on the boxes containing new cell phones?
- If you have noticed, there is a specific warning about not keeping the devices closer to the body than an inch or so; why do you suppose that is?
- Whether one is texting or speaking on the phone, the basic operation is the reception of the signal from a cell tower, and the radiation back from the phone with the information contained. These signals, as all waves do, obey the inverse-square dependence of the signal strength on the distance from the source.
- Even if the exposure to radiation is in the non-ionizing region, whenever your device is far from a cell tower it needs to send more energy back, and so it will radiate more than if it were near one. Interestingly, whenever one is near a cell tower there is more radiation being received by the device since the distance is quite short.
- Do you think it is better to use cell phones in populated cities where there are lots of cell towers, or in rural areas where there will be few and the device may be very far from cell towers?
- When do you think you are exposed to more radiation, in urban centers or rural areas?

The last property of radiation we shall explore is its dependence on angle of incidence. As with distance, there is a change in the amount of radiation absorbed as one changes the angle that the rays or waves make with the surface where the energy is absorbed.

Exploratory Task

Determine the variation and dependence of the amount of light absorbed by a surface on the angle the light makes with it.

Use a light sensor or a light meter downloaded to your phone as an App to measure the light intensity on a chosen smooth surface. Make sure that the distance from the light source to the sensor remains constant for all measurements.

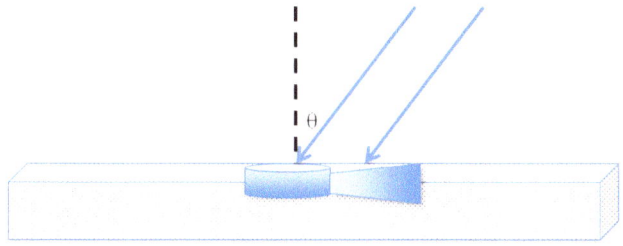

(continued)

Fill in the data table below

Angle (°)	Light Intensity (Lux)

- What will the graph of the relationship between Light Intensity and Angle of incidence look like?
- Reflect on your observations

The next task is a logical extension of the previous activity, and it shows the significance of the relationship between angle of incidence and amount of intensity of light deposited on the surface found above. In addition, the task provides a summary of the effects of both distance from the source, and angle of incidence on the surface, that determine how much light radiation is deposited on a surface.

What is the Cause of the Seasons on the Earth?
Among the most popular ideas investigated with students is their understanding of the reason for the seasons.
Question: what causes the seasons on the Earth?
Answer:
Background
The orbit of the Earth around the sun is basically circular, despite a small variation in distance between the closest point (perihelion), and the farthest one (aphelion) on the orbit. If your answer above was that the distance between the sun and the Earth is the reason for the seasons, this contradicts the fact that a circle has a constant radius; therefore, the distance from the

(continued)

center does not change. Also, consider that during January the Earth is closest to the sun, and yet in the northern hemisphere we experience winter!

Correspondingly, during the summer the Earth is farthest from the sun, and yet we experience summer in the northern hemisphere. By now, you begin to sense that the answer does not depend on the distance, but on the angle that the solar radiation makes with the Earth's surface. This activity demonstrates that property, and should convince those who think the distance is what makes a difference.

Constructing a circular orbit on the floor and indicating four points where a sphere that represents the Earth is placed, we can measure the amount of light radiation that a source of light in the center representing the sun can deposit on the surface of the Earth. The points are perihelion, aphelion, and the two equinoxes (points where the light intensity is the same). But wait a minute, how can we measure differences in light intensity if all these points are at the same distance from the sun?

The answer is that the Earth spins on an axis that is tilted approximately 23.5° to the normal to the plane where the sun and the planets move (the ecliptic). Therefore, the sphere that represents the Earth must be placed at all four points on the orbit, with its axis of rotation or spinning at a *constant value*.

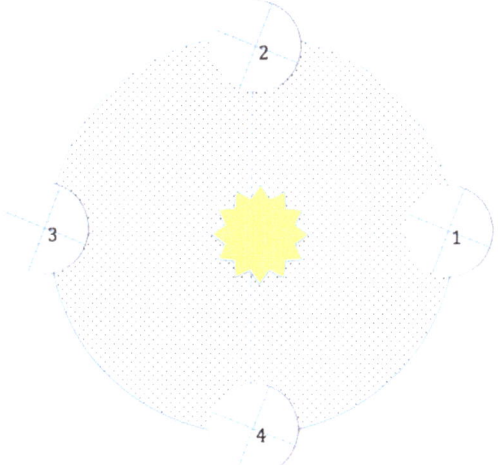

The points 1 (Aphelion), 2 and 4 (Equinoxes), and 3 (Perihelion) are the positions where the sphere representing the Earth is placed, with the axis of rotation making a 23.5° angle with the vertical (dashed line). The placement of the light sensor (or an App such as Lux Camera for the iPhone) shown below is at a point on the Earth that corresponds to this angle.

(continued)

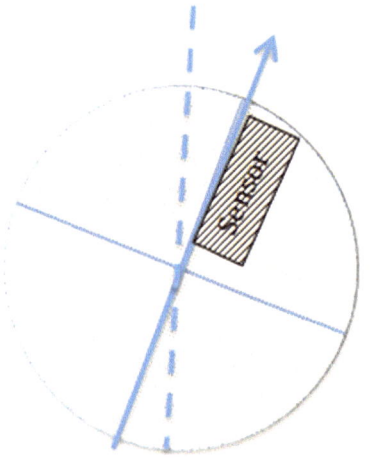

Light Intensity (Lux)	Point 1 (Aphelion)	Point 2 (Equinox)	Point 3 (Perihelion)	Point 4 (Equinox)

Fill in the Table below for the values of light intensity
What point shows the maximum intensity?

9.2   General Properties of Wave Spreading

As we have seen with light and sound, the decrease in intensity as a function of the distance from the source of the waves follows a specific formula, that of an inverse-square dependence on the distance. This property can now be explained in general terms for all wave phenomena, as well as for all energy dispersion.

As Fig. 9.4 shows the energy intensity will be deposited on an area section A at the surface of a sphere surrounding the source. As the distance beyond increases in terms of R the radius of the sphere, the intensity progressively decreases. The change is such that at the shown distances $2R$, $3R$, and $4R$, the area sections where the intensity is deposited will be 4A, 9A, and 16A, respectively. This clearly shows that as the distance increases, the intensity decreases as a function of the distance squared.

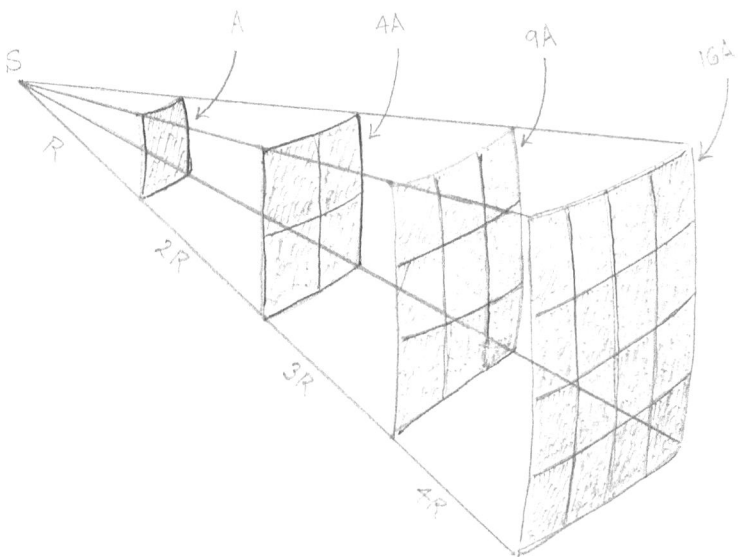

Fig. 9.4 The figure shows the spreading of the area where the energy falls as the distance from the source of the energy increases. If the origin is a point source (*S*) and the intensity of the energy is *I*, then at the surface of the sphere the energy intensity will be $I/4\pi R^2$ where *R* is the radius of the sphere. On the surface of the sphere this energy will be deposited on area *A*. As the distance beyond the sphere increases, the intensity decreases as a function of the distance squared

Chapter 10
Waves and Sensory Perception

We have explored several properties of waves up to this point, particularly those of light and sound while emphasizing the physical characteristics displayed by such wave phenomena. In this chapter we shall deal with some of the physiological, and perhaps even psychological aspects of our daily exposure to waves; after all, throughout history humanity has attempted to understand waves by beginning at our level of perception, and then extrapolating to areas beyond our direct experience with the confidence gained from familiar situations.

One of the features of our observations of natural phenomena is what we consider to be symmetry. The symmetry of an object or a physical system is a property that remains unchanged under certain transformations or changes upon observation. It is a physical or mathematical feature of the system (observed or intrinsic) that is "preserved" under some change. There is some fascinating pre-historical evidence of the role of symmetry in the evolution of the human species.

Similarly to what other species do in dealing with tasks, humans have relied on the use of the hands for the production of tools, and as tools themselves. We can take as a starting point in our discussion of symmetry a pre-historic example. Archaeologists have determined that before a certain time period, which can only be ascertained based on radioactive dating to be roughly between 1.4 and 1.9 million years ago [1], there seemed to be no significant preference for the use of either hand. We can use handedness by humans as a factor that relates to symmetry, in the sense that preferences for either right- or left-hand use show a departure from ambidexterity, or the existence of a 50-50 tendency in the use of the hands.

© Springer International Publishing Switzerland 2017
F. Espinoza, *Wave Motion as Inquiry*, DOI 10.1007/978-3-319-45758-1_10

Exploratory Task

Ideally a person should be ambidextrous, or able to accomplish tasks skill-fully with both hands. However, in the case of writing we all show a tendency to prefer one hand over the other, although some can probably write equally well with either hand.

Determine your skill in completing a tracing task; use the time it takes to completely draw inside various shapes without touching the figure. If you touch it, record the time taken from the start, and then repeat the task.

Procedure

Use the following setup to draw a path inside each shape, first with your preferred hand, and then with the other. Begin at S and move clockwise until you return to the starting position. While you draw, someone else will record the time it takes you to complete the task. Repeat by moving counterclock-wise and determine if there is a difference.

Two different sheets of paper should be used (one for each direction), so as to avoid being influenced by previously erased attempts.

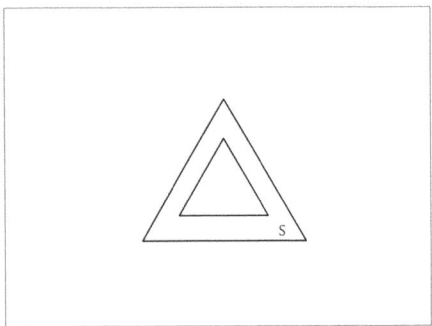

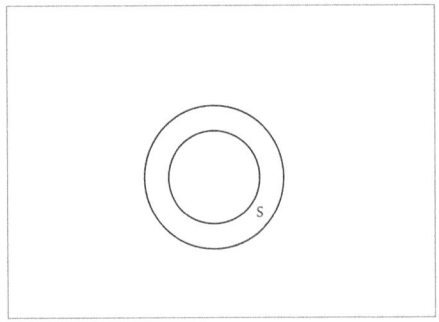

(continued)

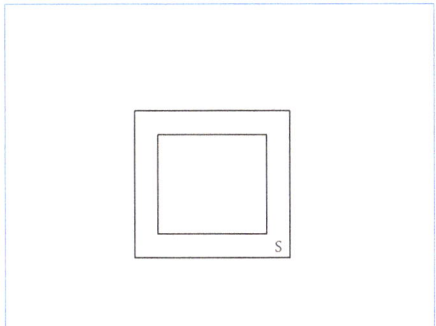

Fill in the table below

Trial	Triangle (right hand)	Circle (right hand)	Square (right hand)	Triangle (left hand)	Circle (left hand)	Square (left hand)
Clockwise direction						
1						
2						
3						
Av.						
Counterclockwise direction						
1						
2						
3						
Av.						

Discuss the results in terms of the time difference in completing the tasks. Is there symmetry (similar times for task completion) between

(A) The left and the right hand?
(B) The clockwise and the counterclockwise direction?

Is there a pattern for the times taken, as you repeat the tasks?
Extension
As an application to writing, perform a similar task available online at: http://legacy.mos.org/sln/Leonardo/LeonardoRighttoLeft.html

Application to Sound

The concept of symmetry can also be applied to our perception of sound in various ways; a particularly simple task is to determine whether your hearing is symmetric in the sense of detecting frequencies by each ear.

To determine if both ears are equally sensitive to frequency, a task can be undertaken where you can test each ear for the maximum audible frequency that can be heard. The task can be done by going online (http://onlinetonegenerator.com/).

When at the website choose "hearing test" and generate the signal while covering one ear (or putting cotton inside it). Determine the maximum value of frequency that is audible with each ear; fill in the table below (Table 10.1).

- Are the mean values equivalent?
- Discuss any differences noticed.

The concept of symmetry can be traced back to Greek origins, expressed as their perception of proportion in many aspects of life. Symmetry evolved from an aesthetic concept to one heavily tied to functionality, as expressed by Roman architecture and engineering. During the Renaissance it became predominantly used in the visual arts.

The most commonly found use of symmetry is in the use of left and right as part of the perception of geometric symmetry. There are other types of symmetry that we can explore:

1. Translational symmetry. It is used when one moves an object a distance without a change in orientation.
2. Rotational symmetry. It consists of turning an object through an angle around an axis as a reference.
3. Reflection symmetry. It is based on specular or regular reflection at a surface, obeying the laws of reflection that we have explored before.
4. Inversion symmetry. It is based on the motion of a point on a system of coordinates, such as the Cartesian one (x, y, and z axes). A point can be moved from + to − on the coordinates, with respect to the origin, or another point.

Table 10.1 Values of maximum audible frequency for each ear

Trial	Left ear	Right ear
1		
2		
3		
Mean		

Translational Symmetry

Translational symmetry results in a pattern with periodicity (it repeats in the same interval of time). Figure 10.1 provides examples of this type of symmetry.

Rotational Symmetry

A figure has a rotational symmetry if the following conditions are met:

(a) There is a point in it that the figure can be turned around a certain number of degrees and still look the same.
(b) The figure appears to be unchanged even after it is rotated.
(c) Its image, after a rotation of less than 360°, appears exactly the same as that of the original figure.

Figure 10.2 provides several examples of rotational symmetry.

Reflection Symmetry

Reflection symmetry is a type in which one half of the object is the mirror image of the other. A figure may have both horizontal and vertical lines of reflection. Figure 10.3 shows examples of reflection symmetry.

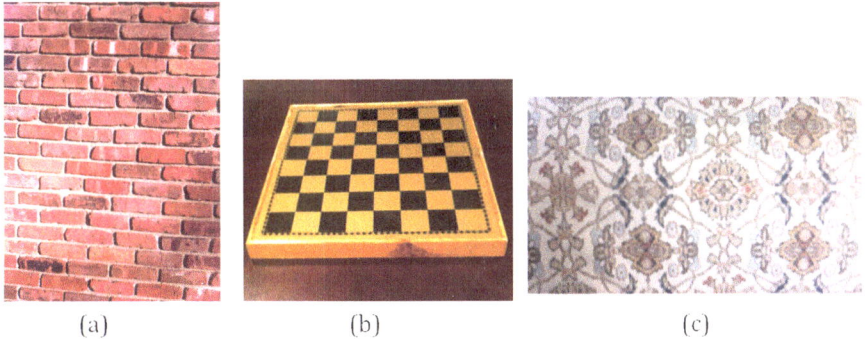

(a)   (b)   (c)

Fig. 10.1 Three examples of translational symmetry are shown. In (**a**) a brick façade shows a repetitive pattern with one preferred direction. In (**b**) a checker or chessboard shows no preferred direction for the pattern. In (**c**) the pattern is more complex, and there are other types of symmetry present, but one can choose a pattern among those that will exhibit translational symmetry

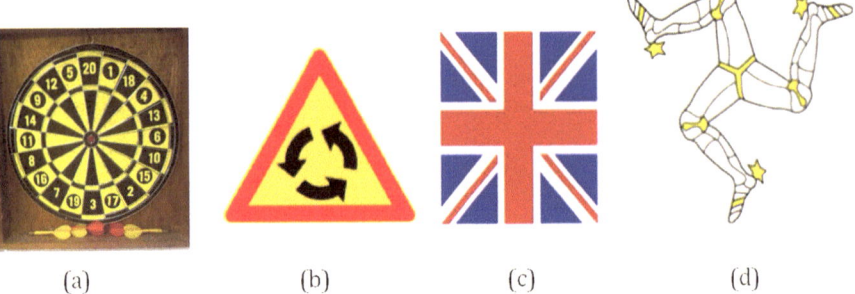

(a)   (b)   (c)   (d)

Fig. 10.2 Four examples of rotational symmetry are shown. In each case you can pick a point and see that the three conditions stated above are met. The exception of course is for the darts in (**a**)

Fig. 10.3 Examples of reflection symmetry; if it weren't for the difference in shading between the two halves in (**a**) the figure would be completely symmetric upon reflection along the *dashed line*. The figure shows vertical reflection symmetry, but not horizontal. If we were to draw a horizontal line cutting the figure in half, there would be no symmetry. In (**b**) there is vertical as well as horizontal reflection symmetry for the figure, but not for the number. In part (**c**) there is both horizontal and vertical symmetry, no matter how we orient the figure

Inversion Symmetry

Inversion symmetry is in a sense a generalized view of rotation, being more accurate than reflection. It can be more challenging to determine due to the larger number of dimensions involved. In two dimensions, a point reflection is the same as a rotation of 180°. In three dimensions, a point reflection can be described as a 180-degree rotation combined with reflection across a plane perpendicular to the axis of rotation.

The difficulty one may experience in seeing the symmetry in Fig. 10.4 is due to the fact that a three dimensional effect is being shown on a two dimensional plane. The following figure is an attempt to explain this by taking a point on a coordinate, and then doing each part of the inversion symmetry separately.

Exploratory Task

The concept of reflection symmetry can be applied to other situations besides looking at figures. We can use an extension of the previous task where the skill in completing the figures was explored, to one where a mirror image of the figure can be used to determine how one's skill in completing the task is affected by reflection.

The following figures show the arrangement of the experimental setup.

The first figure shows the arrangement using one of the shapes (the triangle); the idea is to use the image of the shape by blocking the drawing of the shape itself. In this example a wooden block is used, but any object that properly blocks the drawn shape will suffice.

 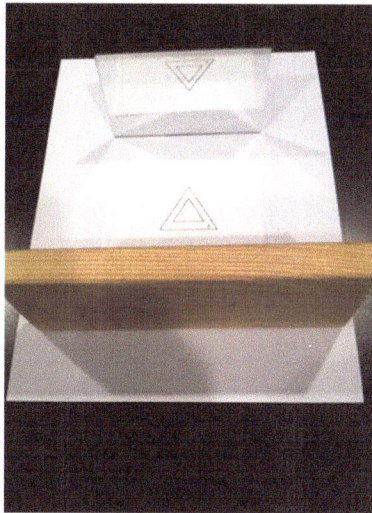

(continued)

The figure with the hand shows the proper view. Make sure that only the reflection of the figure is visible to you, then repeat the tasks previously described. It is expected that the tasks will be more challenging and the times taken to complete the tasks will be much longer. One should only use the dominant hand (the one that you prefer to write with) to determine the number of trials needed to successfully complete the task (remember, to go around without touching the shape).

Fill in the table that follows (the number of rows may need to be modified, depending on how many trials it takes to successfully complete the tasks).

Clockwise				Counterclockwise			
Trial	Triangle	Circle	Square	Trial	Triangle	Circle	Square
1				1			
2				2			
3				3			
Av.				Av.			

- Is there symmetry between clockwise and counterclockwise directions?
- Plot the time taken to successfully complete each figure, as a function of the number of trials needed.
- Are there patterns to the graphs of the data taken to complete the tasks?

We are only exploring the symmetry upon reflection of the direction (clockwise vs counterclockwise) in this task. Further explorations are possible but should be left as extensions or additional projects to explore symmetry.

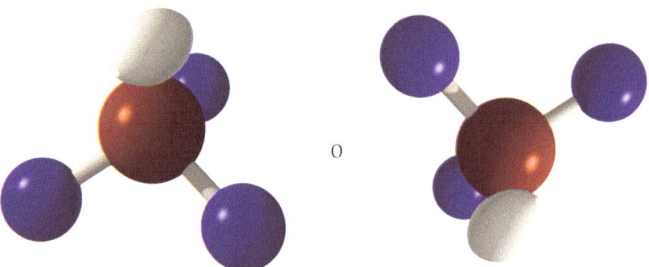

O

Fig. 10.4 An example of inversion symmetry, showing that the molecule appears reflected and inverted at the same time, with respect to the point O. Notice that a simple flipping of the figure alone (the top part of the figure appearing upside down) would not constitute inversion symmetry, since the orientation alone would have changed, and there is no reflection yet

Figure 10.5 shows in detail the process; the shape represents the entire molecule. In part (a) the reflection is shown with respect to the vertical axis, and in part (b) the inversion is shown with respect to the horizontal axis.

Symmetry in Physics

There are a number of important applications of symmetry in physics, chemistry, and biology. One of the most interesting is a property called chirality.

Fig. 10.5 This figure is a detailed explanation of the inversion shown in Fig. 10.4. The shape represents the entire molecule; in part (**a**) the reflection with respect to the vertical axis is shown, whereas in part (**b**) the flipping is shown with respect to the horizontal axis. The result of both operations is inversion

Experimental Task

(continued)

Letters and words can have or lack symmetry; in the enclosed figures, reflection symmetry is exhibited by some letters but not by others. We can clearly see on the figure that X has reflection symmetry but R does not. On the other figure we can see why the word AMBULANCE needs to be written backwards in front of emergency vehicles, so that drivers can see them on their mirrors.

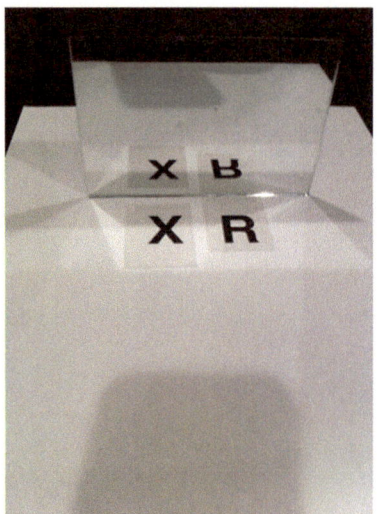

Identify the letters in the figure above that possess the following types of symmetry:

- Rotational symmetry _____
- Horizontal reflection symmetry _____
- Vertical reflection symmetry _____

Chirality: The symmetry of an object determines whether or not it has chirality. A figure or an object is said to be **chiral** (to have **chirality**) if it is not identical to its mirror image or, to put it more precisely, if it cannot be mapped to its mirror image by rotations and translations alone. For example, a right shoe is different from a left shoe, and clockwise is different from counterclockwise.

A molecule is *achiral* (not chiral) when a combination of a rotation and a reflection in a plane, perpendicular to the axis of rotation, results in the same molecule. It may lack some forms of symmetry, but it could have others (such as rotational). Figure 10.6 shows some examples, although we shall not deal with chirality in this text, other than to introduce it as an example of symmetry (or lack thereof) in nature.

Symmetry has been linked to conservation laws through a fairly recent formulation called Noether's theorem, which states that any differentiable symmetry of the action of a physical system has a corresponding conservation law. Translational symmetry = conservation of linear momentum and energy, and rotational symmetry through a fixed angle = conservation of angular momentum.

Geometrical symmetry has been observed for a long time, in terms of the shape of objects. A polygon (a shape with flat or straight sides) has three or more sides. Polygons can be regular (all sides are equal), or irregular (unequal sides). Figure 10.7 shows regular and irregular polygons.

There is an interesting application of symmetrical polygons with a long history. Since the time of the Pythagoreans it has been known that the only regular polygons that can cover a surface without leaving any gaps are: the equilateral triangle, the square, and the regular or equilateral hexagon.

Figure 10.8 shows how the three regular polygons, the equilateral triangle, the square, and the regular hexagon can be used to fill out a space without leaving any gaps in between. The construction of tiles, textiles, and pottery designs show the effective use of such property. Of course, the use by bees of the last pattern in building a honeycomb is intriguing in itself.

Combining other regular polygons and filling in the spaces between them leads to interesting patterns such as tiles. Combining pentagons and hexagons can form a slightly imperfect sphere, although these cannot cover a flat surface. A great example is that of a soccer ball, as shown in Fig. 10.9.

 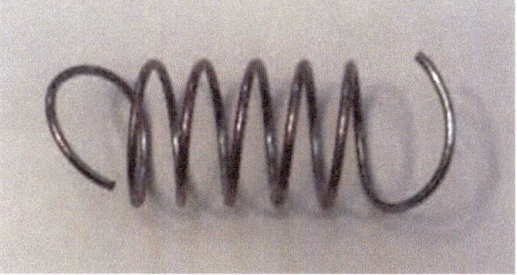

Fig. 10.6 Two examples of objects exhibiting chirality; the first one is exhibited by nature itself, the growth of a sea shelf, while the second (the spring) is human-made

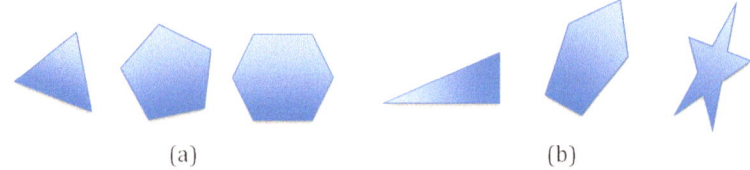

(a)   (b)

Fig. 10.7 The figure shows both types of polygons. In part (**a**) the isosceles triangle, pentagon, and hexagon included are considered regular since their sides are all equal. In part (**b**) the right triangle, unequal pentagon, and star are considered irregular

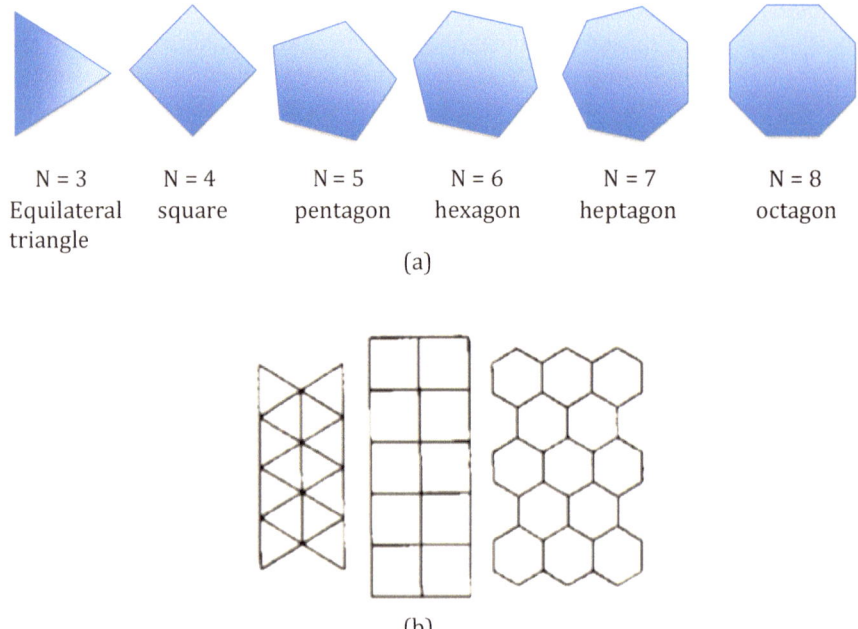

Fig. 10.8 Part (**a**) shows the first six polygons. Part (**b**) shows that the only ones that can fill a space without leaving any gaps are the equilateral triangle, the square, and the regular hexagon

The design of soccer balls has evolved from its very primitive stages to the modern aerodynamic examples seen today. Interestingly enough, there are some characteristics of the motion of these balls through the air that have raised engineering questions, since some irregularities have been observed in their trajectories. It may be interesting to speculate to what extent the basic shape has an impact on this, as shown in Fig. 10.9.

Another area where interesting applications of perceptual features leading to odd observations occur is in the case of illusions. An illusion can be defined as a discrepancy between long-term memory and real-time data as the brain interacts with the environment. Illusions can be classified into three major categories:

1. *Physical* illusions, which are the result of purely external physical processes.
2. *Physiological* illusions, which involve our sensory apparatus as it processes external information.
3. *Psychological* or *cognitive* illusions, which are the result of the internal processing mechanisms within the brain.

Generally speaking, any representation of an object that exists in three dimensions onto a two dimensional surface is in a sense an illusion. Figure 10.10 shows some examples of visual illusions.

Figure 10.11 shows a particular type of illusion. When the collection of irregular but identical shapes is arranged on the left, the pattern can be interpreted as either a white grid made by irregular lines against a black background, or a collection of identical "black squares." The pattern is the same on the right, but

Fig. 10.9 A soccer ball does not have a perfectly round shape since it is made of both regular and irregular polygons, added together in the way it is manufactured. Notice that there is a five sided (*pentagon*) and a six-sided (*hexagon*) put together, so the result is that the sphere obtained by doing this is not perfectly round

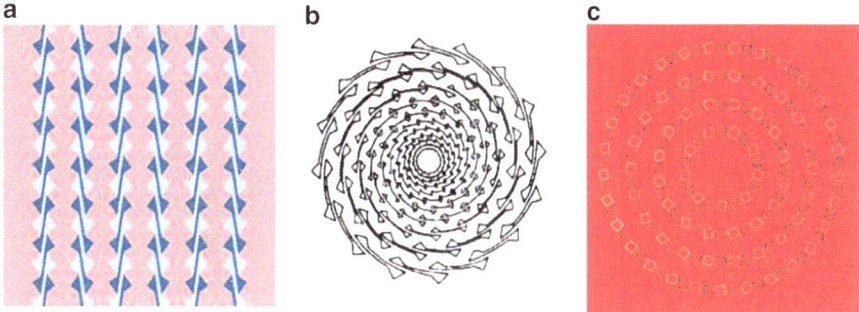

Fig. 10.10 Some examples of visual illusions. In reality, *the lines* in part (**a**) are all parallel, but the perception is that they are slanted due to the other shapes in the figure. In parts (**b**) and (**c**) *the concentric circles* by definition do not intersect, but the perception is that they spiral inwardly in both cases, regardless of the background

Exploratory Task

Describe your observations of the following figure.

instead the separations between the black squares or the lines of the white grid are straight. In the latter case one sees an additional feature of the pattern, dots in between.

Another interesting illusion is that resulting from using an object called the Necker cube. We can see the illusion emerge by beginning with a two dimensional object, a regular hexagon. Figure 10.12 illustrates the process in detail.

As Fig. 10.12 shows, if one begins with a regular hexagon and then adds a corner the resulting shape is a cube, whose depth can be perceived but not clearly determined. Adding a second corner completes the illusion, which is called the Necker cube. The illusion consists of the change in orientation of the cube where the front face repeatedly seems to come in and out of the field of view.

Fig. 10.11 The situation described in the second picture is commonly referred to as the Hermann grid. It is the result of an illusion that arises from a transformation in the shape of the lines

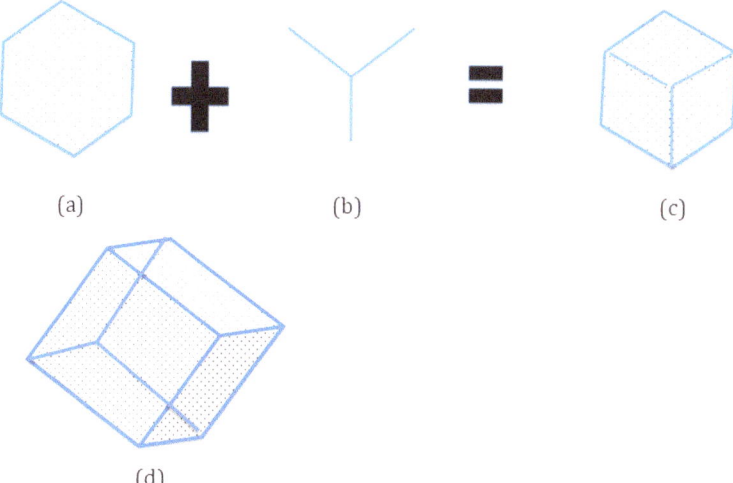

(a)   (b)   (c)

(d)

Fig. 10.12 The Necker cube can be generated with a regular hexagon (**a**) and then adding a corner (**b**); the result (**c**) is a cube whose depth can be sensed but not clearly determined. The beginning of the illusion depends on whether the corner is said to be either convex or concave (pointing in or out of the plane). Adding a second corner gives rise to the full illusion in (**d**), which is that the orientation of the cube alternates

Exploratory Task

The "default" view in the stack of cubes is that the point O is at the origin of the coordinates, and the axes appear as coming out of the plane. Determine how long you can keep viewing the figure showing the stack of cubes so that the point O represents the bottom of the stack, and the axes appear to go into the plane.

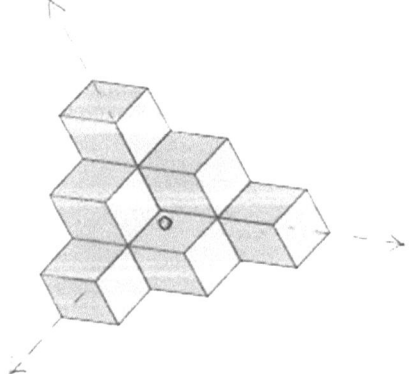

Reference

1. McManus, I. C. (2002). Right Hand, Left Hand: The Origins of Asymmetry in Brains, Bodies, Atoms and Cultures. London, UK/Cambridge, MA: Weidenfeld and Nicolson/Harvard University Press

Chapter 11
Forensic Applications

In this chapter, we shall explore a number of ways in which the properties exhibited by light and sound can be used in forensic investigations. We begin by recalling the various features that waves display, such as reflection, refraction, diffraction, polarization, and intensity variations.

This may require going back to previous chapters to review some of these ideas; in any case, the objective of the chapter is to provide a context for these tasks, so that they are more meaningful in their representation of actual criminal investigations.

The tasks are not arranged in any particular sequence or order, except categorizing them as applications of either the properties of light or those of sound, which incidentally are the same, but they require the use of different equipment.

Applications to Sound

(I)
Use of a motion detector to determine the location and the shape of objects

Context

Imagine that a crime has been committed where the authorities suspect a safe taken from a mansion that could not be opened had to be buried by the thieves. You are given several locations to search for it, and you wish to determine where to excavate. Since you need to take into consideration the time and effort involved in the excavation, you want to have a good idea of where the safe would most likely be, by investigating the sites with ultrasound reflection.

(continued)

The original version of this chapter was revised. An erratum to this chapter can be found at http://dx.doi.org/10.1007/978-3-319-45758-1_13

F. Espinoza, *Wave Motion as Inquiry*, DOI 10.1007/978-3-319-45758-1_11

The following activity illustrates how sound is used to detect objects either underground or underwater, by bouncing sound waves off the objects.

The Motion Detector used in this experiment emits short bursts of ultrasonic sound waves; these waves fill a cone-shaped area off the axis of the centerline of the beam. The Motion Detector allows one to measure how long it takes for the ultrasonic waves to travel a distance to an object and then back to the detector. Using this time and the speed of sound in air, the distance to the nearest object is determined. By reporting the distance to the closest object that produces a sufficiently strong echo, the motion detector can pick up objects in the cone of ultrasound. It is therefore extremely important to avoid placing any other objects near the cone of sound, so that the signals don't confound the data.

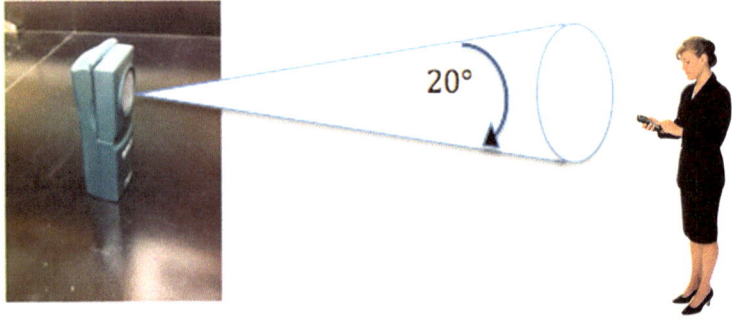

The motion detector sends out an ultrasound cone of approximately 20°. The size of the area (the base of the cone) depends on the distance from the detector multiplied by this angle. Since the motion detector acts as a *transducer* (a device that both emits and receives energy) by functioning as a microphone and as a loudspeaker, there is a minimum distance from the detector that the investigated object should be placed.

The manufacturer of the detector states that this distance should be approximately 40 cm, to account for the existence of a "blind spot" where the signal emitted by the transducer operating as a loudspeaker stops, and the signal obtained by it operating as a microphone begins.

Sound can be used as a probe to measure the size and shape of objects in the atmosphere (in air) and in other substances such as water. The use of sound in water is generally called "sonar." The fact that when a wave meets a boundary between two materials part of the wave is reflected and part is transmitted, allows for the use of sound in diagnostic applications as seen in a previous chapter, as well as in other areas of research. Among the many uses of sound by scientists, archeologists, marine geologists, and oceanographers, employ it to investigate objects underwater. A signal is sent out and bounces

(continued)

back from a submerged surface. Scientists use the speed of sound in water and the time it takes for the signal to bounce back to calculate the depth of the object. A particular advantage in this case is the fact that the speed of sound in water is almost five times greater than in air.

In this activity we'll use the reflection of sound with a motion detector to determine the size (volume) of an object; the measurements can then be compared with the actual dimensions of the object to test the accuracy of the method.

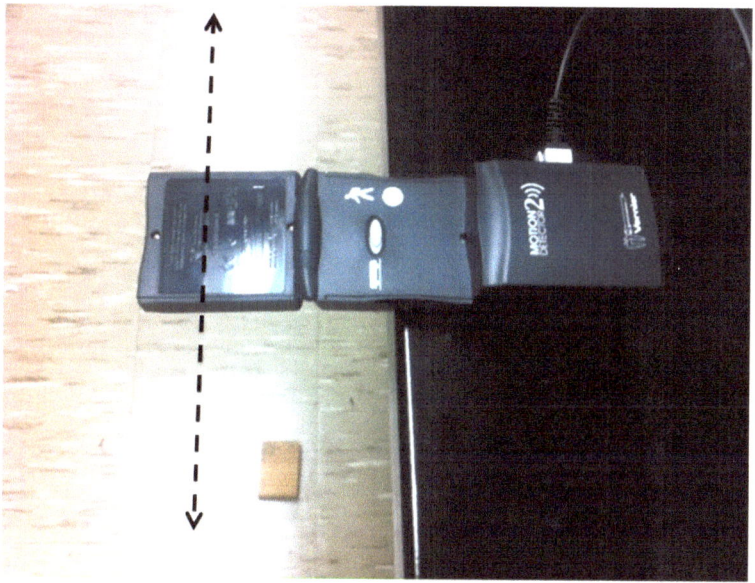

A motion detector is placed on a table facing down so that the ultrasound cone goes over the object (a wooden block) on the floor. One should begin collecting data before encountering the object and continue after having passed it.

The motion detector needs to be moved at a constant speed across the table (along the dashed black line) so that the span occupied by the object on the floor is covered in the amount of time chosen to collect the data.

The graphs obtained may look different, depending on how the reflected signal is interpreted by the motion detector. Either display will yield the particular distance from the flat lines (before and after bouncing off the object). In both cases one needs to subtract the depth or height of the signal from the flat line(s) that represent the floor, or reference level.

(continued)

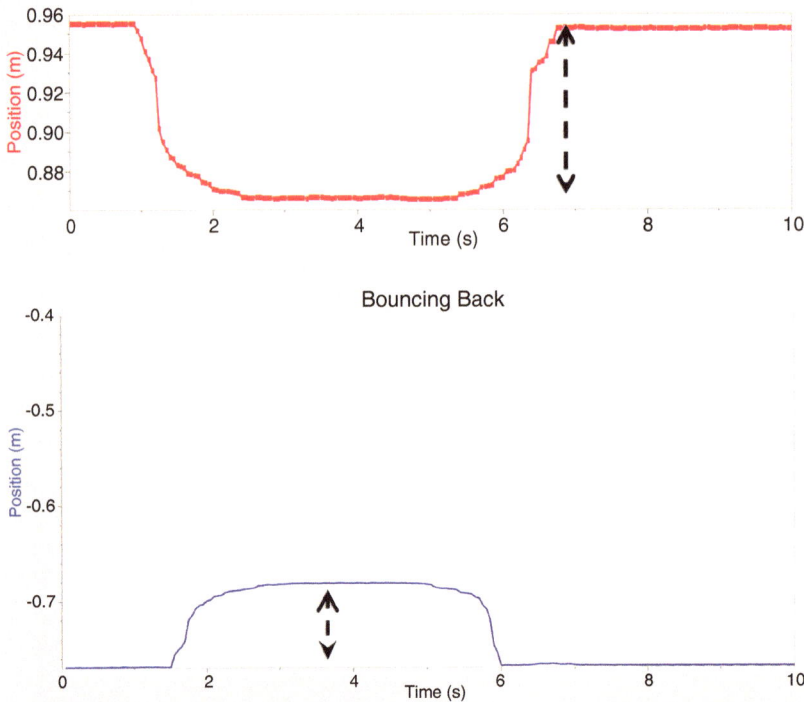

To determine the volume of a cube, each dimension is measured by placing the cube on each one of its three sides, and then collecting the data that represent the particular dimension (length, width, and height).

Note: To find the length, width, and height of each object subtract the distance to the object from the distance to the floor.

Object I

Distance to the floor			
Trial	Length	Width	Height
(1)			
(2)			
(3)			
Average			

Object II

Trial	Length	Width	Height
(1)			
(2)			

(continued)

Trial	Length	Width	Height
(3)			
Average			

Use the average values to fill in the following table:

Measurement	Length	Width	Height	Volume	%error
Object (I) with sound					
Object (I) actual size					
Object (II) with sound					
Object (II) actual size					

$$\% \text{ error} = \frac{\left|\left[\text{Measured Volume}\left(\text{actual}\right) - \text{Estimated Volume}\left(\text{sound}\right)\right]\right|}{\text{Measured Volume}\left(\text{actual}\right)} \times 100$$

Reflections

1. Which was your best result in parts (I) and (II) of the experiment? Why do you think it was better than your other result?
2. In part (II) what object gave the best result in terms of the lowest % error?
3. What are the main sources of error?

(II)
Use of a motion detector to determine the distance from objects and to describe their motion.

Motion detectors are commonly used in a variety of security systems. One of the most effective methods of describing an object's motion is from graphs of position, velocity, and acceleration vs. time. From such a graphical representation, it is possible to determine in what direction an object is going, how fast it is moving, how far it traveled, and whether it is speeding up or slowing down. In this experiment, you will use a Motion Detector to determine this information by plotting a real time graph of *your* motion as you move across the classroom. Let's consider the following scenario.

Context [1]

(Copyright 2016 by AACE and the Education & Information Technology Digital Library (EdITLib), www.editlib.org, included here by permission)

Suppose an intruder accesses a secured area and your job is to investigate the evidence provided by two motion detectors that monitored the room in question. There are two entrances that are located perpendicular (at right angles) to each other and that are monitored by the detectors. The following figure illustrates the situation.

(continued)

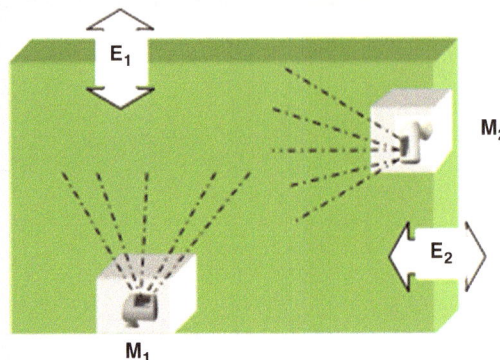

An intruder setting. There are two entrances E_1 and E_2, and the room is monitored by two motion detectors M_1 and M_2

(a) Using the graph below determine which motion detector captured the motion of the intruder, the detector facing the entrance used, or the one perpendicular to it?

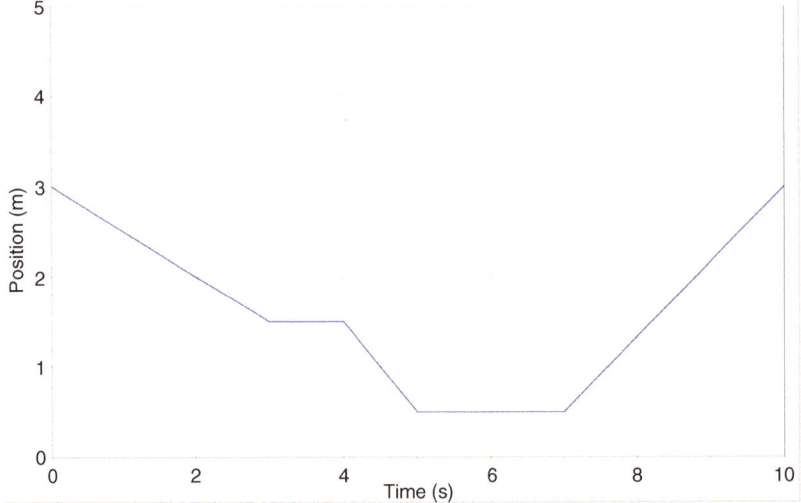

Answer:

_____

(b) How did the intruder move, as shown by the graph of the recorded motion?
   Procedure:
   The Motion Detector measures the time it takes for a high frequency sound pulse to travel from the detector to an object and back. Using this round-trip time and the speed of sound, you can determine the position to the object.

(continued)

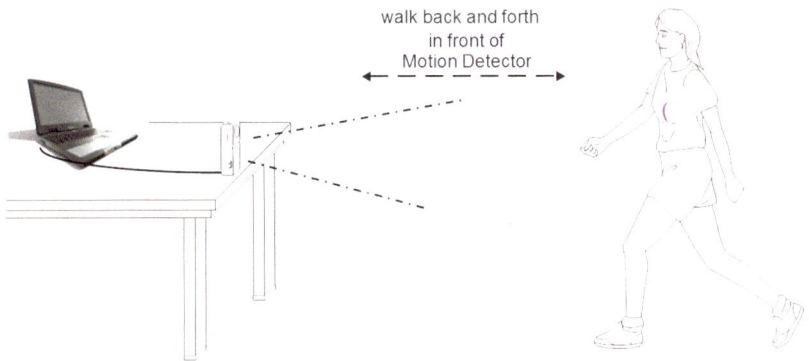

Describe the following graph, and then match it with your own motion (**I**)

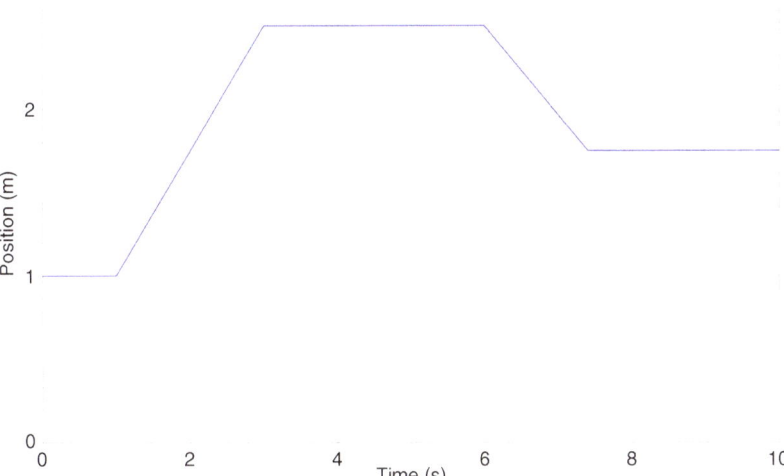

Description:

| Motion Detector | Interface | |

(continued)

Open Logger Pro and go to File, then Open, and then choose Physics with Vernier: 01b Graph Matching (Draw your best motion in the figure below).

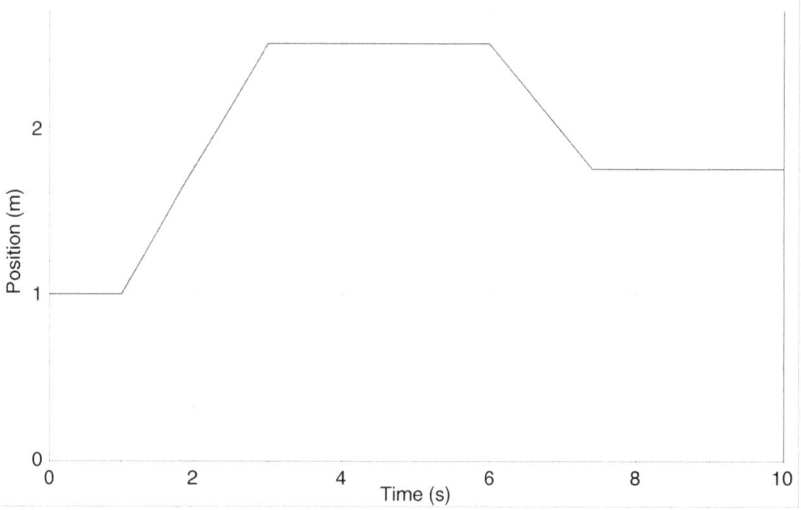

(**II**) Now that you have experienced the effect of your own motion, describe once again the intruder case graph.

Description:

Open Logger Pro and go to File, then Open, and then choose Physics with Vernier: 01c Graph Matching (Draw your best motion in the figure below).

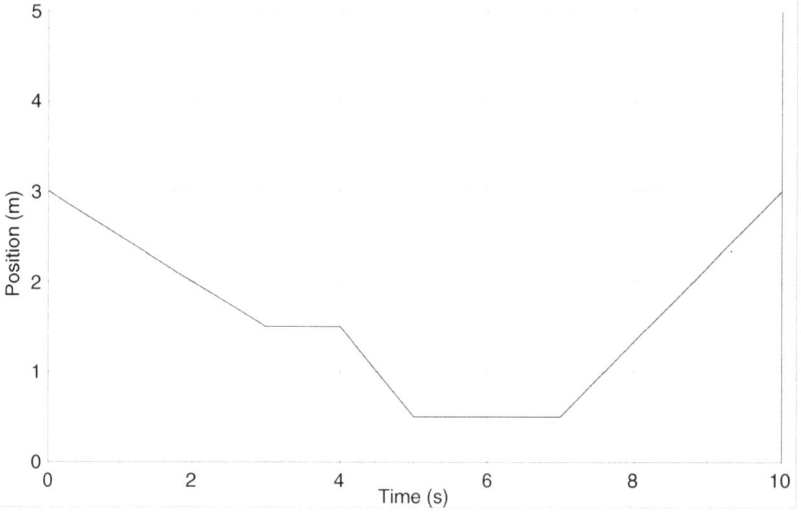

(continued)

Conclusions and Reflections: What have you learned as a result of the tasks where you relate your own motion to its description?

(III)

Use of a microphone to determine the frequency of sounds and other related properties

Context

Suppose a situation involving another safe, where the thieves actually succeeded in opening it at the site of the robbery. The safe has a computer lock similar to a telephone keypad. Each time a number on the pad is pushed, a specific tone is produced. Apparently someone other than the owner of the safe had access to the combination. At this time, the main suspect is the butler, and investigators found sophisticated sound-recording equipment in his apartment. Upon searching his computer hard drive, they discovered files containing waveform patterns. The investigators believe that the butler recorded the sounds made by the safe's keypad and used them to determine the combination of the lock.

The waveforms found in the suspect's computer are included. In each case the wave frequency can be obtained to compare it with the pitch of the sounds produced by the safe as the respective sequence of numbers that opens it. Each waveform is plotted for 3/100 s (0.03 s), and we need to determine the number of cycles (waves) in this amount of time. Dividing 0.03 s by the number of cycles gives the period (T), and then $f = 1/T$ gives the frequency in Hertz. The first three waveforms (a, b, c) are shown with the cycles identified as the pattern that is repeated by the uneven signals. The remaining ones are straightforward in their appearance so the number of cycles can be easily counted.

As an example, the first one (a) is

0.03 s/3 cycles, and so $T = 0.01$ s, and $f = 1/0.01$ s $= 100$ Hertz (Hz)

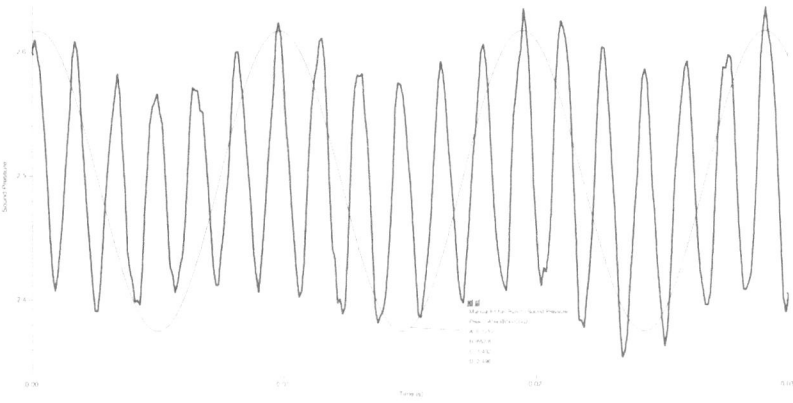

(a)

(continued)

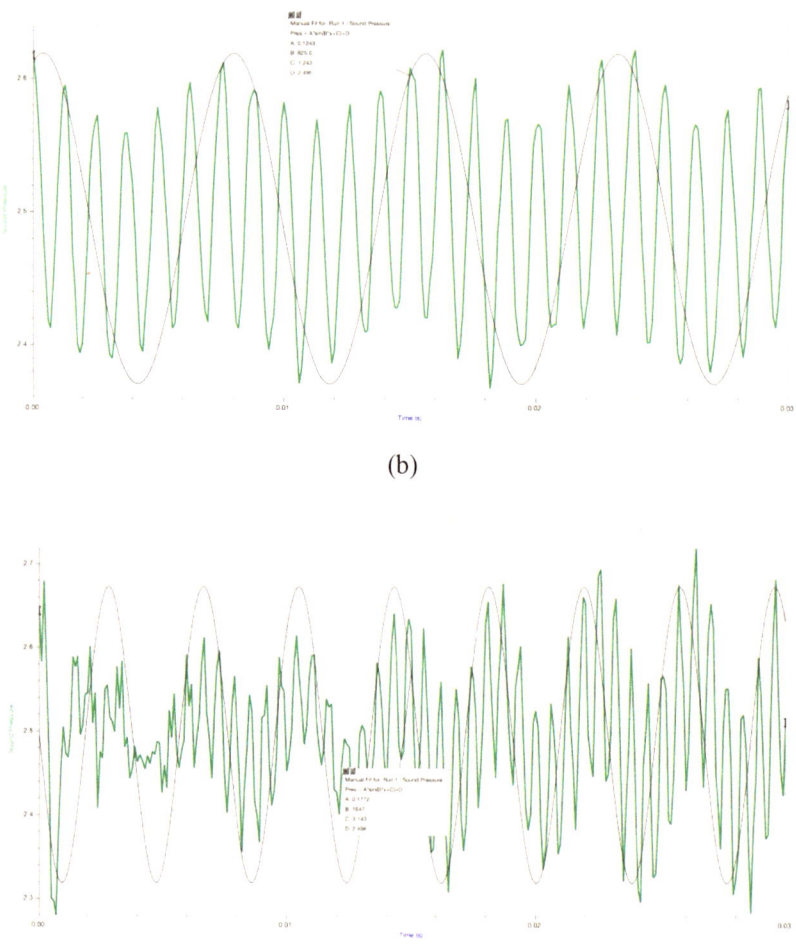

(b)

(c)

(continued)

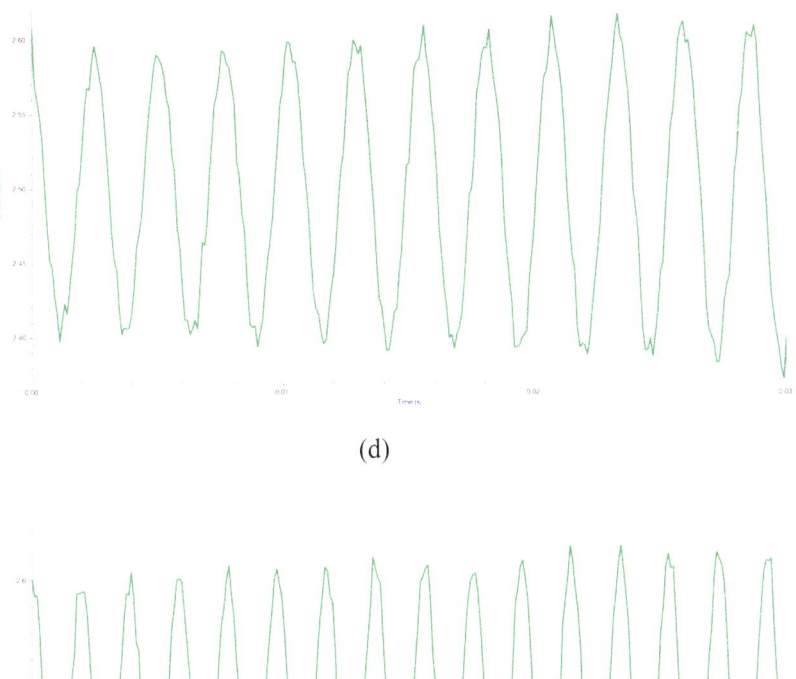

(d)

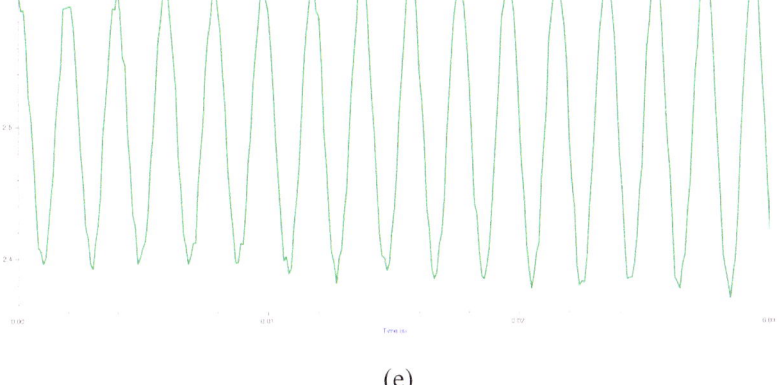

(e)

(continued)

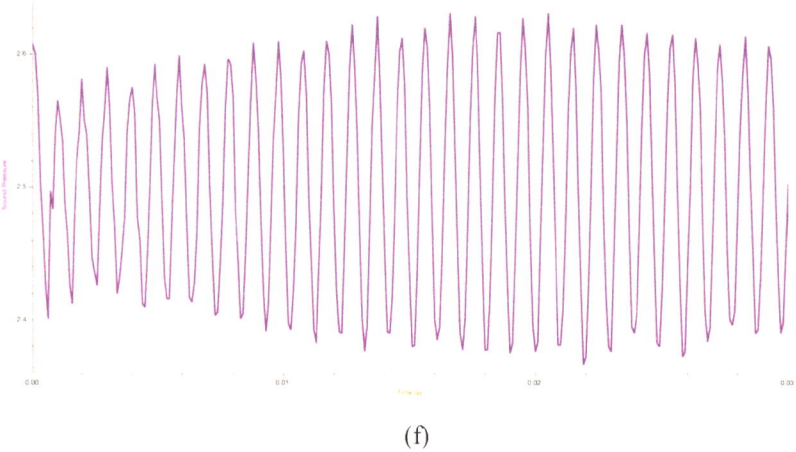

(f)

Waveform	T (s)	F (Hz)
(b)		
(c)		
(d)		
(e)		
(f)		

The investigators determined that the safe could be opened by the sequence formed by the following notes: C_5, C_4, and C_6. Does this pattern appear in the suspect's waveforms?

Applications to Light

(I) Human Reaction Time (Visual)
Dropping a ruler to determine human reaction time

Human reaction time is how long it takes the eyes to tell the brain that the ruler is falling and how long it takes the brain to tell the fingers to catch it. We can use the distance the ruler falls before one catches it to figure out reaction time. We use a kinematic equation (a solution to Newton's second law of motion) to find the height

$$d = d_0 + V_0 t + 1/2 g t^2 \tag{11.1}$$

In this formula, d equals the distance the ruler falls, d_0 is an initial height, V_0 is the initial velocity of the ruler, g equals the local gravitational acceleration (9.8 m/s²), and t is the time the ruler falls.

(continued)

Since the ruler is initially dropped from rest ($V_0=0$, and $d_0=0$ for simplicity), Eq. (11.1) reduces to

$$d = \frac{1}{2} g t^2 \qquad\qquad (11.2)$$

Solving for $t = \sqrt{2d / g}$

Collect data for several trials of dropping the ruler and determining the distance it falls, and then use the average distance to find the reaction time from the equation.

How does the answer compare to the average human reaction time of 250 ms?

(II)

Context

Another piece of evidence against the butler is being used; it consists of a record of light intensity during the time the safe was opened. The detector records the light as a variable pattern (a sine curve) during the entire monitoring process. The butler in his defense submits data from his cell phone that he claims was accidentally left on a table in a room with fluorescent lights with his light sensor app running, and the time during which the signal was recorded shows it as a constant signal. He argues that he could not have been at the room where the signal was captured as a variable pattern. Is he telling the truth?

The variable pattern of fluorescent lights can be determined by using a light sensor whose sensitivity is comparable to that of the human eye. It is generally known that the eye will perceive continuous or fluid motion if the frequency with which static frames (such as in cartoons and films) exceeds a range of values. However, it is agreed that most people will not see any flickering past 100 Hz or frames per second. Since the human eye cannot distinguish between flashes that occur more than about 50 times a second, the light appears to be on all the time.

Investigating fluorescent lights' flickering effect and vision

We make use of a reverse procedure to the one where the eye begins to see a flickering effect of light and then gradually sees the light as a continuous pattern. We shall use a light sensor to determine the emergence of a variable pattern of light intensity from a fluorescent light source, as we change the sampling rate of the sensor, beginning with values where the light is recorded as a constant value. The collection rate of the sensor is changed and light is collected, until a variable pattern appears, and then with it we determine the frequency of the oscillations that give rise to the illusion that the light is continuous.

In this part you will point the light sensor at a single fluorescent light and record its intensity for a very short period of time. The resulting plot of intensity versus time is interesting because it shows that fluorescent lights do not

(continued)

stay on continuously but rather flicker off and on very rapidly. Since the human eye cannot distinguish between flashes that occur more than about 50 times a second, the light appears to be on all the time. The data you collect will be used to determine the period and frequency at which the light flickers.

We use a light sensor connected to an interface and a computer running a program such as Vernier's Logger Pro.

1. When the sensor is connected to the interface and that to a computer a default screen appears with a graph, a table, and a reading box.
2. From the Data Collection icon

choose duration 0.05 s, and sampling rate 20 samples per second.

3. Hold the light sensor near a fluorescent light.
4. Collect data of light intensity versus time for the light source.
5. Fill in the table below for each of the values of the sampling rate and the sensor reading.

Trial	Sampling rate	Light intensity (Lux)
1	20	
2	60	
3	100	
4	200	
5	400	
6	800	
7	1000	

(continued)

6. For what sampling rate did the light intensity begin to look like a broken line?
7. For the last sampling rate value, extend the range of the vertical axis so that the amplitude of the signal is clearly visible, and from the taskbar menu choose:

   Analyze, then **Curve Fit**, then **Sine**
8. From the wave pattern determine the period from the wave form as before, and then the frequency

   Frequency (f) =
   Reflections

Reference

1. Espinoza, F. "Graphical Representations and the Perception of Motion: Integrating Isomorphism Through Kinesthesia into Physics Instruction". *Journal of Computers in Mathematics and Science Teaching* (2015). Volume 34, Issue 2, (pp. 133-154).

Chapter 12
Technological Applications

Applications to Light

One of the traditional challenges in representing reality that humans have encountered is in the reproduction of objects that exist in three dimensions onto a two dimensional surface. Distortions are inevitable, and the attempts have always been to faithfully copy the main features of objects. Since the time of the early Egyptian civilizations, human figures appeared facing sideways on the depictions of daily activity, as well as ritualistic representations.

The difficulty experienced by those individuals attempting to copy a real scene onto a flat surface was eventually resolved during the renaissance. The invention of perspective painting enabled artists to copy reality to a previously unprecedented degree of accuracy. This involved a clever trick of geometrical representation.

Imagine the tricks that can be employed in deliberately including features in drawings that defy logic, in the sense that they appear to display features that we would never experience in the world of three dimensions that is our everyday physical setting. Figure 12.1 shows two examples of impossible objects that appear possible as represented initially; however, upon further inspection one can see that they are impossible, given our reference of everyday experience.

The interesting aspect of Fig. 12.1 is that despite being impossible to construct in reality, they appear as they do by a trick that enables the representation to go from two to three dimensions. There are other ways to accomplish a representation in three dimensions from a figure that is constructed in two dimensions. Figure 12.2 illustrates two ways in which this can be done.

Another way to determine depth is to use the method of parallax, where a shift in the angle of view between two objects that are closer to the observer is larger than the shift between them if they are farther away. This is illustrated in Fig. 12.3.

The solution to the problem of representing objects in three dimensions on a two dimensional surface that ancient artists had faced was to come from an idea developed by the Arabs and introduced to European artists in the art of painting. Figures 12.4 and 12.5 demonstrate its use, which is based on a perspective based on

© Springer International Publishing Switzerland 2017
F. Espinoza, *Wave Motion as Inquiry*, DOI 10.1007/978-3-319-45758-1_12

Fig. 12.1 Two examples of impossible figures are shown. In both cases, initially they appear correctly represented, until further inspection reveals that they cannot possibly be built/constructed to look as they appear

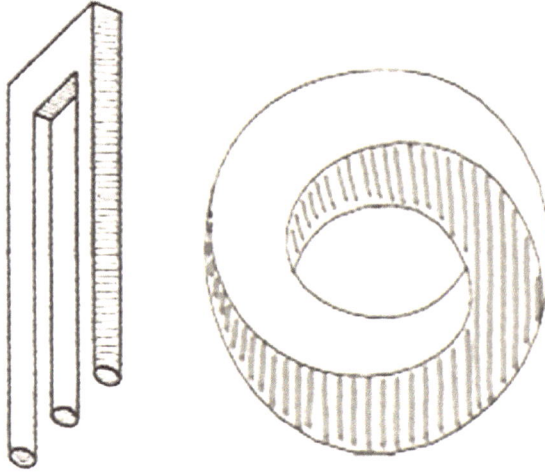

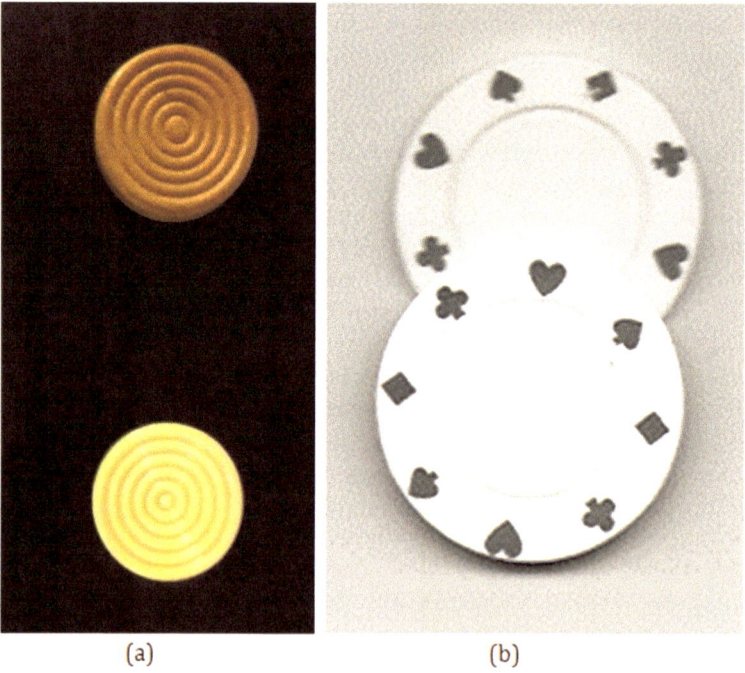

(a)   (b)

Fig. 12.2 The figure shows two ways to create the illusion of depth, or to convey a three dimensional view from a two dimensional representation. In (**a**) two objects of different size are placed next to each other, the lower one seems farther away than the top one. In (**b**) two objects of identical size are placed next to each other, but one overlaps the other, thus making the partially blocked object to seem farther away

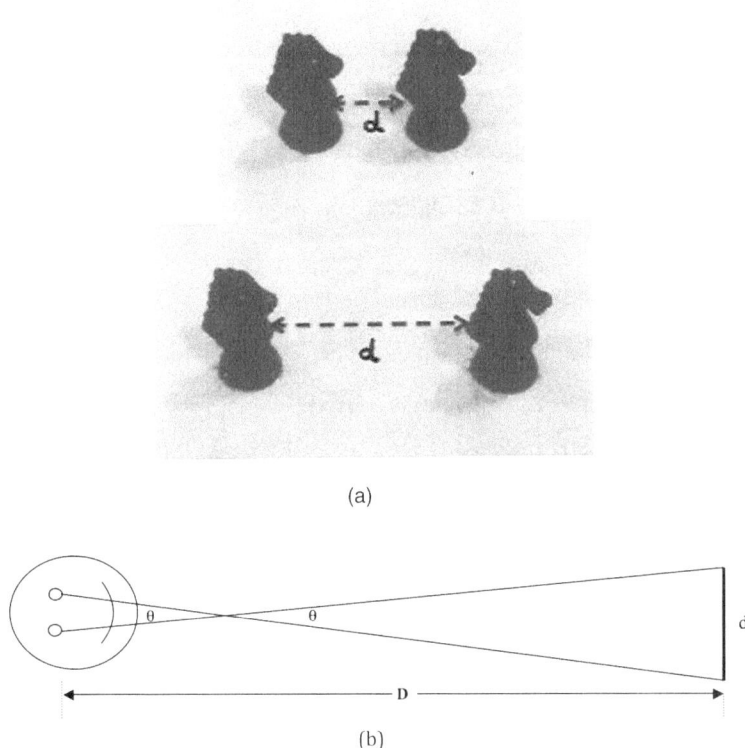

(a)

(b)

Fig. 12.3 The phenomenon of parallax consists of the perception of the distance D that an object is from the observer, as depending on the angle θ that is formed by the separation d between the two chess pieces shown. In (**a**) the separation d is different; the shorter d is, the farther away they seem from the observer. In (**b**) as the separation d between the two pieces changes, the angle θ also changes, and so based on that observation the two objects can be determined to be farther (larger D) or closer (smaller D) to the observer

what the eye sees by projecting the reflected rays of light from objects onto the eye as a cone with its base representing the size of the object.

Figure 12.4 illustrates the concept of centric rays that was invented by the Arab thinker Al Hazen [1]. He envisioned rays coming from objects and converging onto the eye as being of two types: (1) those that came in at an angle, or obliquely and not entering the eye since they bounced off the sides of the visual cone repeatedly, and (2) those called centric by virtue of entering the cone straight or along the center and successfully converging on the eye.

This idea proved to be enormously important in allowing artists during the renaissance to create the illusion of depth in paintings, as demonstrated in Fig. 12.5.

Figure 12.5 shows how one can project a flat surface using another one but slightly distorted, to create a sense of depth. Part (b) shows how one can extend the projection to a point, which enables the creation of a multitude of such points to reproduce an entire image. The advantage is that the image now appears to have

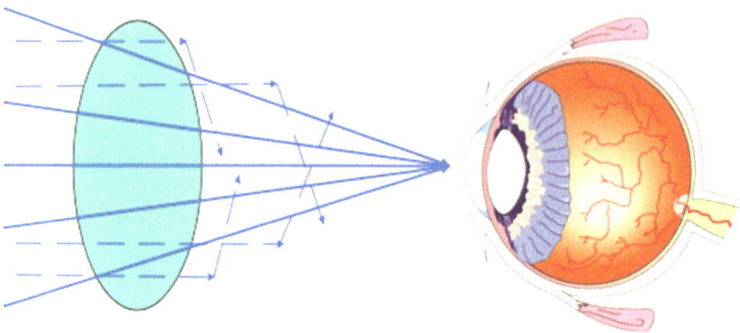

Fig. 12.4 The diagram shows the basic idea behind the visual cone. The eye is imagined to perceive the size of objects as a cone surrounding them, where the base would change in size depending on their distance from the eye. *The dashed lines* represent the rays that enter the base of the cone but strike the sides of the cone and bounce off without reaching the eye. *The solid lines* represent the centric rays, which successfully reach the eye

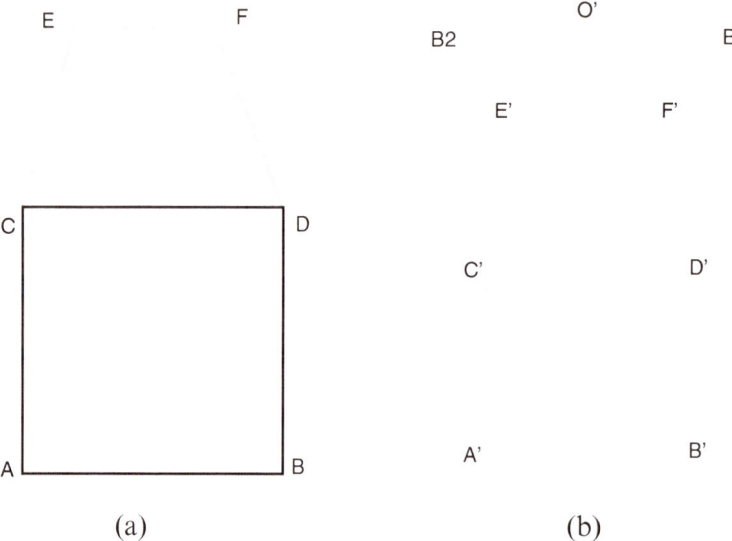

(a)   (b)

Fig. 12.5 The square face ABCD can be projected by the use of the distorted CDEF. This creates the illusion of depth on the plane of the paper or surface on which it is drawn

depth, beginning with an object in two dimensions that has none. This process constitutes the basis of perspective geometry.

The creation of images in three dimensions from those on a surface using properties of waves is based on reflection and the interference of transverse waves. The technique is generally known as interferometry, and it can be used with a variety of light sources. In the production of holograms a laser beam is sent from a source and encounters a beam splitter (e.g., a half mirror) that sends it in two different directions.

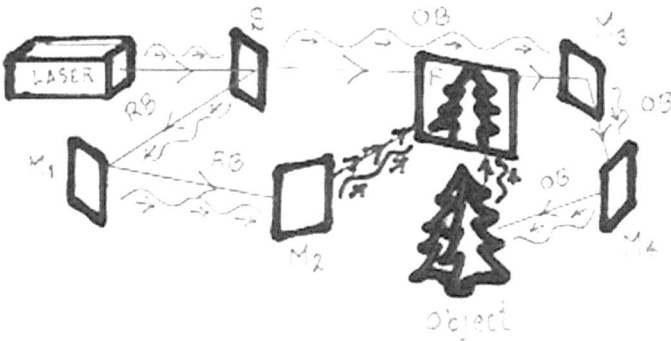

Fig. 12.6 The arrangement shows how a hologram is produced. The light beam is depicted both as a ray and as a wave; the beam sent from the laser is split by (S) into the object beam (OB) and the reference beam (RB). Each beam reflects twice from mirrors in this case; RB then strikes a photographic film (F), while OB bounces off the object and then strikes the film. These last two beams are shown highlighted in the diagram, as they converge to produce the three dimensional image of the object on the film

These new beams reflect from mirrors, one beam is made to bounce off the object to be projected onto a photographic film, and converge on the film along with the other beam. The arrangement is illustrated in Fig. 12.6.

An interferometer is a device that enables a very accurate determination of distance, and it has a wide variety of applications. Figure 12.7 shows a schematic where a laser beam is made to split, the two beams then reflect from two mirrors, and then recombine to converge on a detector.

The function of the interferometer needs to be explained in detail, since it is the *phase difference* between interfering waves that is the central feature in its widespread utility and popularity as a device to measure distances very accurately.

The discussion about how waves combine in Chap. 6 in terms of constructive and destructive interference can also be carried out in terms of another property of waves called their *phase*. This property was deliberately left out of those listed in Chap. 2, and it can be appropriately introduced now.

We can refer to Fig. 1.4a to illustrate mathematically what Hooke's law enables us to do. When describing the behavior of a spring with a mass attached at its end, and then set in motion, the force described by Hooke's Law is the net or total force acting on the spring. When we apply Newton's Second Law ($F = ma$, along the x direction), the sum of the forces is expressed as

$$\Sigma F_x = -kx = ma_x \qquad (12.1)$$

where F_x is the force, k is the spring constant, x is the amount the spring is stretched or compressed, m is the mass attached to the spring, and a_x is the acceleration along the x-axis.

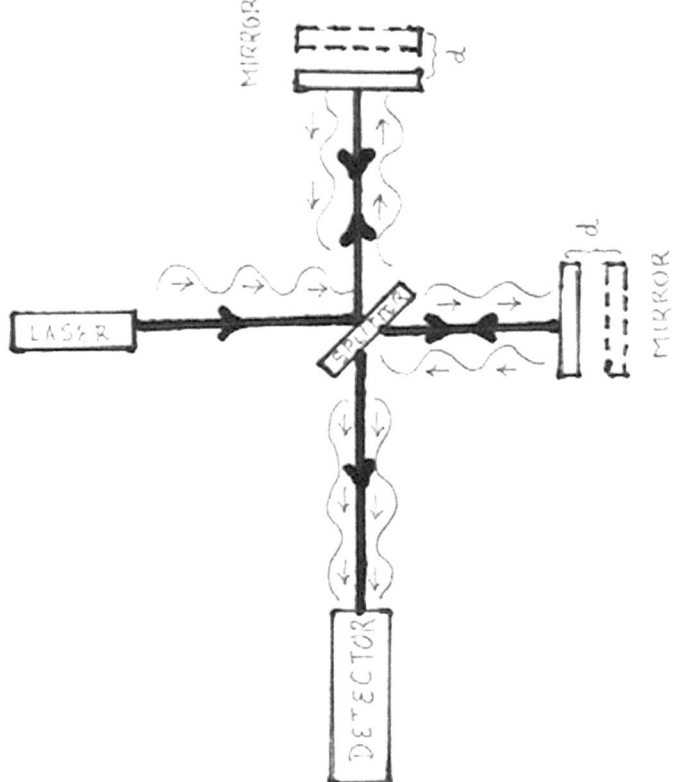

Fig. 12.7 An interferometer consists of a laser beam split into two beams; each beam reflects off a mirror, these are then recombined at the splitter again and converge on a detector. The phase difference between the reflected beams gives a measure of the distance d that either mirror has shifted in its position. All beams are depicted as both rays and waves, along with their directions. The splitter serves both functions, to refract and split the original beam, and then to recombine the reflected beams

Rearranging Eq. (12.1) we get

$$ma_x + kx = 0 \qquad\qquad (12.2)$$

Equation (12.2) is known as a differential equation, whose solutions are algebraic equations. In this case the acceleration a_x is the second derivative of x with respect to time. You can understand this relationship by extrapolating from the solution to an algebraic equation (which is a numerical one). In other words, a numerical answer is a solution to an algebraic equation, whereas an algebraic equation is a solution to a differential equation.

The sine and cosine functions are commonly used as solutions to Eq. (12.2) that represent the sinusoidal pattern described by the motion of the spring. These can be expressed as either

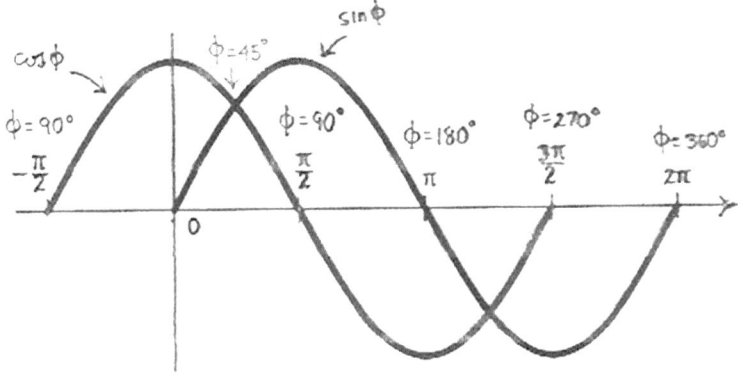

Fig. 12.8 The diagram shows that the trigonometric functions sine and cosine have a constant phase difference of 90° or $\pi/2$. If $t=0$ (at the origin), sine begins with zero amplitude, but cosine is at its maximum. The phase difference of 45° corresponds to 1/8 of a wavelength, 90° to ¼ of a wavelength, and so on. The *shaded region* represents a complete cycle for the cosine function, but not for the sine, since it lags by that constant phase difference

$$x(t) = A\cos(2\pi ft + \Phi) \text{ or } x(t) = A\sin(2\pi ft + \Phi) \qquad (12.3)$$

where A is the amplitude, f is the frequency, 2π expresses the fact that both functions sine and cosine repeat themselves after a complete cycle, and Φ is the phase.

The solution that we wish to use is $x(t) = A \cos(2\pi f t + \Phi)$ since it satisfies the following conditions:

When $t=0$, the spring is stretched to the maximum value of x (A the amplitude), and its velocity is 0.

These two trigonometric functions have a *constant* phase difference of 90°, as illustrated in Fig. 12.8.

Exploratory Tasks

(I) Can you keep track of the values for the cosine function, given those for the sine function? Use Fig. 12.8 to fill in the table

Position	Sine	Cosine
1	45	
2	90	
3	180	
4	270	
5	360	

(continued)

(II) Following the time evolution of sine and cosine functions

Use the online simulation "Fourier: Making Waves" available at https://phet.colorado.edu/en/simulation/legacy/fourier

Make sure the default frame looks like the following figure

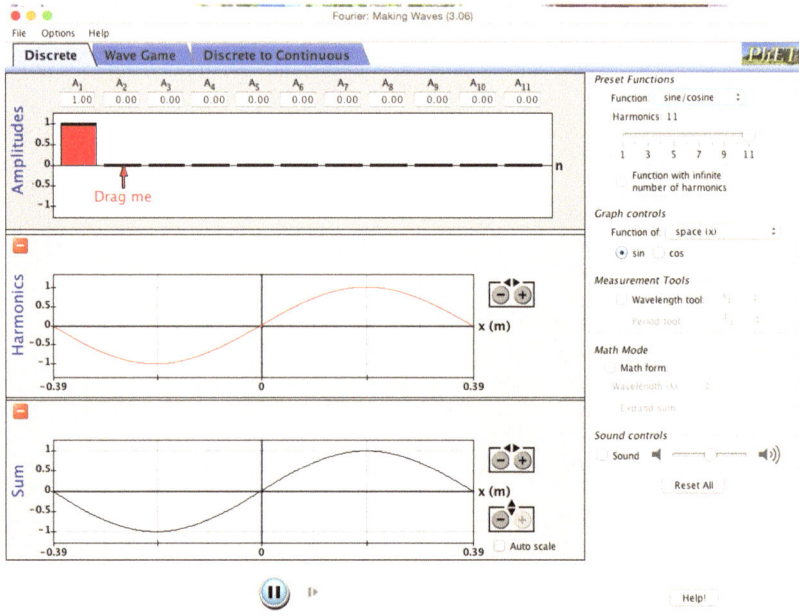

Minimize the Harmonics graph so that the simulation looks like this

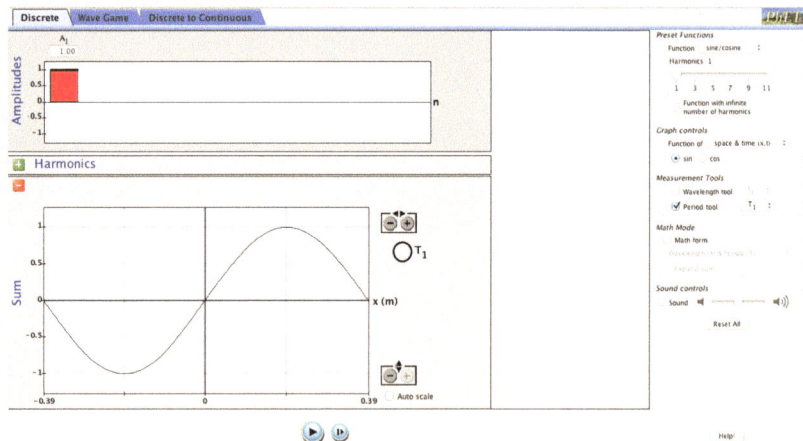

Press the Play button and observe how the sine curve moves to the right; then change the Graph control function to cosine (cos) and repeat. What do you notice on the vertical axis with each function as they begin to move?

In addition, sine and cosine functions have an additive property that enables us to obtain more complex patterns that in turn represent many properties of waves, including constructive and destructive interference.

A particularly important technique for studying properties of waves is known as Fourier synthesis (when waves are added to construct a complex pattern), or as Fourier analysis (when a complex wave is decomposed into its corresponding parts). In both cases we can see that using just sine and cosine functions results into an immense variety of wave applications.

Exploratory Task

Using the same PhEt simulation as in the previous task, choose "Square" from the preset functions button. The screen will now look like the following figure, and the additional information is included to show how the individual sine waves combine to produce the complex pattern that is a square wave. The top graph shows the amplitudes of each of the harmonics (recall that these are multiples of the fundamental frequency), in this case multiples/fractions of A_1 the amplitude of the fundamental. The middle graph shows how the harmonics add up as waves, and the bottom graph shows the net or total result (the Sum of all these waves).

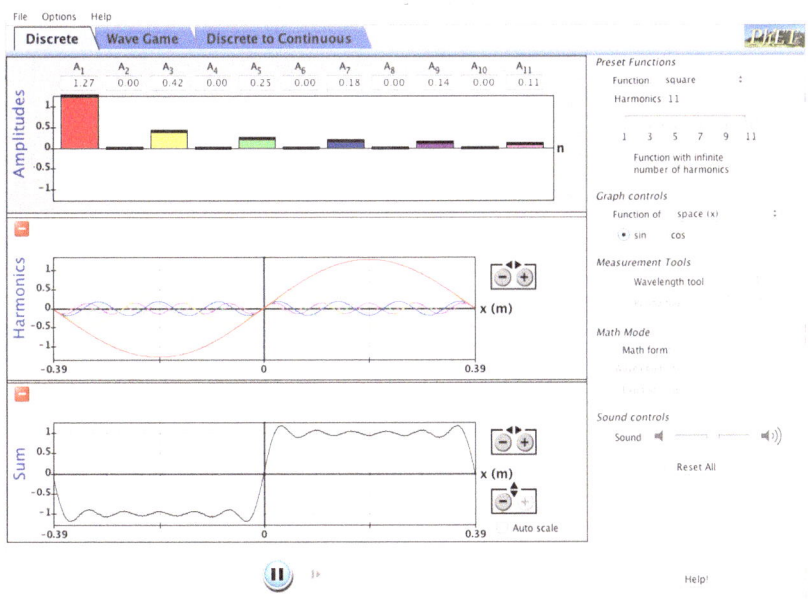

Click the Sound button and analyze the patterns in the Harmonics and the Sum graphs; Move the rider on the Harmonics bar from 11 to 10 and listen to the tone.

(continued)

(1) What happens to the tone as the number of harmonics is reduced from 11 to 1?
(2) What happens to the patterns of both the Sum and the Harmonics graphs as the rider is moved from 11 to 1?

Figure 12.9 shows the result of two waves in phase; the resulting or total amplitude will be greater than that of either wave alone.

Interferometers will use the shift in the patterns that results when waves interfere, depending on whether or not there is a phase difference between them. As we saw in Chap. 7, the interference of waves results in alternating bright and dark regions on a screen. Figure 12.10 illustrates the pattern produced by the interference between two waves.

When the beam that has been split and later reconstructed from the reflected beams is analyzed in the detector, the pattern of fringes produced is compared to the original one, which has been represented by Fig. 12.10. Several likely outcomes are illustrated in Fig. 12.11.

The figure shows the details of what different outcomes would look like. In the top diagram of Part (a) the corresponding incident and reflected waves are in phase, or they differ by a whole wavelength; in this case the lack of a shift in the fringe pattern is used as evidence that the mirror has not moved. The middle and bottom diagrams in both parts (a) and (b) illustrate that shifts in the fringe patterns are an indication that the mirror has moved.

There have been several historically important experiments where the results of a lack of a shift in the fringe pattern (the Michelson Morley experiment), or the evident shift (the recent evidence used by the LIGO scientific collaboration), have led to very important discoveries. In the case of the Michelson Morley experiment, the constancy of the speed of light was experimentally established, in addition to the refutation of the idea of the existence of the ether (a medium for electromagnetic waves to propagate through).

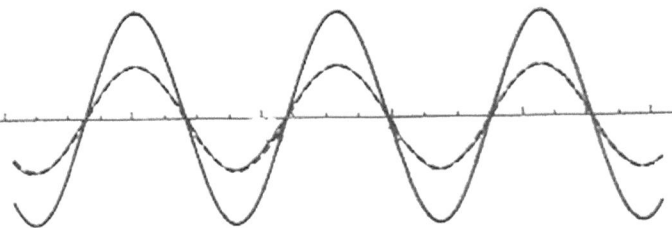

Fig. 12.9 Illustrating the additive property of waves. The waves shown have different amplitudes, but since they are in phase they reinforce one another, resulting in a combined wave with amplitude that is greater than the individual ones

Fig. 12.10 The illustration
is meant to represent the
regions of constructive
(*dark bands*) and
destructive (*light bands*)
interference produced by
two waves

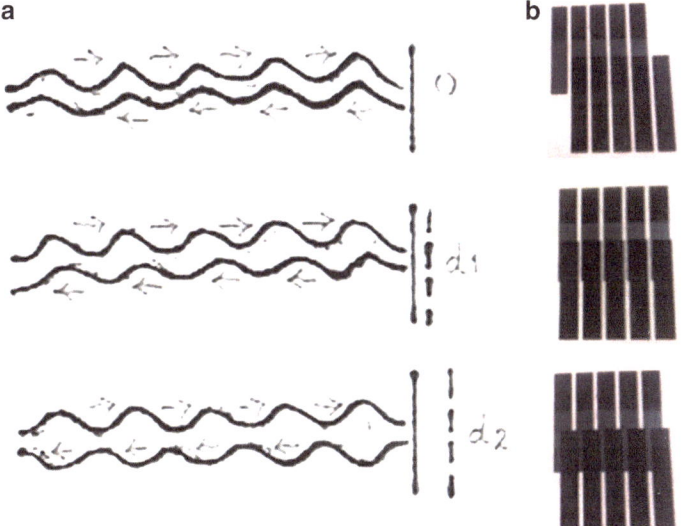

Fig. 12.11 The diagrams are meant to illustrate three outcomes. Part (**a**) shows the patterns produced by a mirror where the incident and reflected waves are either in phase or out of phase by different amounts. Part (**b**) shows the accompanying interference patterns. If they are in phase the mirror will not have moved; however, if they are out of phase the mirror must have shifted either by an amount d_1 or d_2

The very recent announcement of the discovery of gravitational waves [2] also rests upon evidence provided by the observation of a shift in the fringe pattern of signals. In this case, test masses were used instead of mirrors to determine the gravitational force between them. The advanced LIGO detector made use of a modified Michelson interferometer.

Exploratory Task
Use Fig. 12.11 to answer the following:

If the wavelength of a laser beam is 632.8 nm, remember that
1 nm = 1.0×10^{-9} m, and the fringe shift on the right is determined to be in the
middle and bottom parts ¼ λ and ½ λ, respectively, what are the corresponding distances that the mirror will have shifted?

Applications to Sound

One of the interesting applications to sound has its basis on the phenomenon of
acoustic levitation, which is the use of sound to balance the pull of gravity on a
mass, or its weight. This has been accomplished by creating a focus point in space
using standing waves. Recall that a standing wave results whenever a single wave
interacts with itself, or two or more waves interact with each other. The result is a
set of alternating regions of maximum and minimum constructive interference.

In our discussion of standing waves produced by air vibrations there were alternative descriptions of the nodes and antinodes that constitute the maxima and
minima, depending on whether we used the air displacements or the pressure
variations.

If we look at Fig. 5.3b in detail we can point out the specific ways in which the
air displacements and pressure variations are related.

Figure 12.6 is just Fig. 5.3b reused to illustrate the relationship between air displacements and pressure variations.

As Fig. 12.6 illustrates, the antinodes at the end of the air vibrations inside a tube
that is open at both ends represent regions where the air is displaced the maximum.
At this point we need to remember that what causes the air vibrations in the first
place is the exertion of a force that is transmitted to the air as pressure. Consequently,
if we use pressure variations instead, the antinodes correspond to regions of minimum
air pressure, that is, atmospheric pressure.

At the same time, everywhere inside the tube, the nodes represent regions of minimum air displacement or maximum air pressure.

This is what has been used to develop acoustic levitation applications. A source
of sound such as a loudspeaker creates a sound that is reflected off a smooth surface
across from it, or two loudspeakers create sounds opposite each other. In the first
instance, the reflection of the sound creates a standing wave in the space between
the loudspeaker and the reflecting surface, whereas in the second case the standing
wave is produced by the interactions between the two oppositely sent sounds.

By locating the regions of maximum air pressure in between, and furthermore by
finding a focus, or point where the energy transmitted is the greatest, small objects
can be placed there and made to levitate. That is the basic idea behind acoustic
levitation, as has been recently demonstrated in a number of ways.

Conceptual Task

View the video on acoustic levitation at
http://www.youtube.com/watch?v=odJxJRAxdFU
After finishing the video answer the following:

- Why do you suppose they use four speakers instead of two, which could also create a focus in two dimensions?
- Where are the objects suspended in the standing wave produced by the four speakers?
- How many different materials did you observe being levitated?

Discuss three different ways that you can imagine acoustic levitation being used for.

Quantitative Tasks

1. What is the range of the size of the objects levitated in the video (in mm)?
2. Where are the objects located in a standing wave in two dimensions?
3. Draw your answer to question 3
4. If we use the size of the largest object levitated, according to your diagram is this distance the wavelength?
5. If we take the speed of sound to be 340 m/s and the frequency of 40 kHz they used, what is the wavelength?
6. How does the wavelength from question 5 compare to the size of the larg-est object levitated?
7. Explain your result in question 6.

References

1. Espinoza F. (2012). *The Nature of Science*. Rowman & Littefield publishers.
2. Abbott et al., (2016). Observation of Gravitational Waves from a Binary Black Hole Merger. *Physical Review Letters*, 116 (1-16).

ERRATUM

Wave Motion as Inquiry

Fernando Espinoza

© Springer International Publishing Switzerland 2017
F. Espinoza, *Wave Motion as Inquiry*, DOI 10.1007/978-3-319-45758-1

DOI 10.1007/978-3-319-45758-1_13

In the original version of chapter 8, the equation in page 151 was incorrectly captured. It should read as follows:

$$\left(V - V_s\right) f_0 = V f_s$$

In the original version of chapter 11, the equation for the % error in page 203 was incorrectly captured. It should read as follows:

$$\% \text{ error} = \frac{\left|\left[\text{Measured Volume (actual)} - \text{Estimated Volume (sound)}\right]\right|}{\text{Measured Volume (actual)}} \times 100$$

The online version of the original chapters can be found at
http://dx.doi.org/10.1007/978-3-319-45758-1_8
http://dx.doi.org/10.1007/978-3-319-45758-1_11

Index

A

Absorption, 59, 74, 139–142, 175, 177
Abstraction, 7
Accuracy, 7, 8, 10, 19, 21, 23, 24, 156,
    201, 215
Al Hazen, 217
Ambidexterity, 183
Amplitude, 1, 18, 28, 34–36, 38, 39, 45–49,
    76, 104–106, 108–110, 114, 123, 149,
    162, 213, 221, 223, 224
Anagrams, 9
Analyzer, 146, 147
Anatomical, 3, 149
Applications, 1, 3, 6, 12, 42–55, 59, 67, 68,
    72–74, 80, 81, 83, 86–101, 111–114,
    116, 117, 121, 125, 132–135, 138, 149,
    157–158, 185–187, 191, 193, 194,
    199–204, 206, 207, 210–213, 215–217,
    219, 221, 223–226
Astronomers, 14, 158

B

Basilar, 158–160
Biorhythms, 6
Birefringent, 141
Broadcasting, 8

C

Charge, 6, 167, 173, 174
Chirality, 191, 193
Circular, 19, 30, 33, 96, 99, 119–121, 123,
    155, 156, 178, 179
Circumference, 19, 20, 155
Collision, 2, 103, 174

Component, 27, 28, 30, 32, 33, 41, 42, 46,
    141, 149, 173
Compressional, 32
Congestion, 32
Constructs, 6, 68, 72, 81, 84, 96, 104, 106,
    119–121, 123, 133, 134, 155, 156, 179,
    193, 215, 216, 219, 223, 225, 226
Cycles, 1, 3, 6, 33, 34, 51, 207, 221

D

Damping, 49
Density, 6, 29, 41
Didactic, 1, 7
Diffraction, 117, 119–130, 132–134, 144, 199
Discrepancy, 14, 194
Dispersion, 180
Displacement, 8–12, 30, 33, 34, 39, 47, 104,
    111, 112, 226
Distance, 2, 3, 7, 9, 14–17, 30, 34, 40, 56, 57,
    61, 63, 64, 69–72, 83, 88, 89, 93, 100,
    111, 119, 120, 125–127, 129, 130, 142,
    143, 145, 149, 154, 155, 159, 164, 167,
    168, 170–173, 177–181, 186, 200–203,
    210, 211, 217–220, 226, 227
Disturbance, 6, 8, 9, 27, 28, 30, 31
Doppler effect, 149–152, 154, 155, 158
Duration, 5, 14, 165, 212

E

Earth, 12, 14, 15, 141, 143, 178, 179
Earthquake, 6, 31, 32, 48, 159, 160
Echo, 15, 17, 73, 100, 151, 200
Eclipse, 14, 45
Elastic, 9, 15

© Springer International Publishing Switzerland 2017
F. Espinoza, *Wave Motion as Inquiry*, DOI 10.1007/978-3-319-45758-1